SOCIETY FOR EXPERIMENTAL BIOLOGY

SEMINAR SERIES · 3

ANALYTICAL AND QUANTITATIVE METHODS IN MICROSCOPY

ANALYTICAL AND QUANTITATIVE METHODS IN MICROSCOPY

Edited by

G. A. MEEK

Senior Lecturer in Electron Microscopy
University of Sheffield

AND

H. Y. ELDER

Senior Lecturer in Physiology
University of Glasgow

CAMBRIDGE UNIVERSITY PRESS
CAMBRIDGE
LONDON · NEW YORK · MELBOURNE

Published by the Syndics of the Cambridge University Press
The Pitt Building, Trumpington Street, Cambridge CB2 1RP
Bentley House, 200 Euston Road, London NW1 2DB
32 East 57th Street, New York, NY 10022, USA
296 Beaconsfield Parade, Middle Park, Melbourne 3206, Australia

First published 1977

Printed in Great Britain at the
University Press, Cambridge

Library of Congress Cataloguing in Publication Data

Main entry under title:
Analytical and quantitative methods in microscopy

(Seminar series – Society for Experimental Biology; 3)

Includes index
1. Microscope and microscopy – Technique – Congresses. I. Meek, Geoffrey, A. II. Elder, Hugh Y. III. Series: Society for Experimental Biology (Gt. Brit.). Seminar series – Society for Experimental Biology; 3
QH207.A5 578′.4 76-22983

ISBN 0 521 21404 1 hard covers
ISBN 0 521 29141 0 paperback

CONTENTS

CONTRIBUTORS

Appleton, T. C., Physiological Laboratory, Downing Street, Cambridge CB2 3EG, UK.

Bradbury, S., Department of Human Anatomy, University of Oxford, South Parks Road, Oxford OX1 3QX, UK.

Burge, R. E., Physics Department, Queen Elizabeth College, University of London, Campden Hill Road, London W8 7HH, UK.

Chandler, J. A., Tenovus Institute for Cancer Research, Welsh National School of Medicine, Heath, Cardiff CF4 4XX, UK.

Elder, H. Y., Institute of Physiology, University of Glasgow, Glasgow G12 8QQ, UK.

Ferrier, R. P., Department of Natural Philosophy, University of Glasgow, Glasgow G12 8QQ, UK.

Goldstein, D. G., Department of Human Biology and Anatomy, University of Sheffield, 3 Clarkehouse Road, Sheffield S10 2LA, UK.

Hutchinson, T. E., Department of Chemical Engineering and Materials Science, University of Minnesota, Minneapolis, Minnesota 55455, USA.

Horne, R. W., John Innes Institute, Colney Lane, Norwich NR4 7UH, UK.

James, N. T., Department of Human Biology and Anatomy, University of Sheffield, 3 Clarkehouse Road, Sheffield S10 2LA, UK.

Meek, G. A., Department of Human Biology and Anatomy, University of Sheffield, 3 Clarkehouse Road, Sheffield S10 2LA, UK.

Ploem, J. S., Department of Cytochemistry, University of Leiden, Sylvius Laboratory, Wassenaarseweg 72, Leiden, The Netherlands.

PREFACE

Recent advances in microscopy have to a large extent centred around the application of analytical methods, both physical and mathematical, to established techniques of microscopy in order to obtain accurate quantitative information. These new techniques of image analysis frequently involve the use of highly sophisticated apparatus, the higher realm of mathematics, or both; spheres of activity in which the pure biologist traditionally lacks expertise. An SEB symposium was therefore organised with the specific aim of explaining to the research biologist the principles, applications and shortcomings of a selection of the more promising of these novel techniques, and to give some idea of the cost and availability of commercial apparatus and the interpretation of the sort of results which may be expected from typical biological specimens.

The organisers of the symposium hope that the information given in detail by the distinguished contributors to this volume may be helpful to those engaged in biological research. The cost and complexity of the majority of the latest techniques of microanalysis (the simplicity and cheapness of the manual techniques of stereology described by Dr James being a notable and shining exception!) puts them out of reach of the majority of biologists, especially those working in universities. However, the necessary apparatus may be at hand – in another more physics-orientated department perhaps – and, armed with a certain amount of comprehension of the problems involved and knowledge of how they might be solved, one can but ask if one may use someone else's apparatus to solve one's own problem. We hope that the necessary knowledge may be found in this book.

June 1976

G.A.M.
H.Y.E.

G. A. MEEK

Introduction

The basic aim of all forms of microscopy is to obtain as much information as possible from a sample of matter (the specimen) by interacting it with different forms of radiation (electromagnetic or particulate) in different modes (transmission or reflection; flood-beam or scanning-beam). Until quite recently the microscopist has been content with a qualitative, or at best a semi-quantitative, description of the specimen based on visual information.

Attempts to obtain numerical data from typical two-dimensional images of microscopical sections are not new. In the past, areas of objects on positive prints have been measured with planimeters, and optical densities of objects on photographic negatives have been measured with microdensitometers. These simple techniques, though direct, involve much tedious labour in the gathering of relatively little data. Manual methods for the numerical analysis of micrographs, both light and electron, have been greatly simplified by the fairly recent introduction of the technique of 'point counting', one of the basic techniques of stereology. Stereology (a term coined as recently as 1963) is basically a very simple, accurate and powerful technique for extrapolating from two dimensions to three; from areas measured on micrographs of sections to calculated volumes present in the solid specimen from which the section was taken. The principles and applications of point and intersect counting and stereology in general are explained in the first article in this book, contributed by Dr Nigel James of Sheffield.

Stereology is applicable to the analysis of conventional photographic microscope images, both light and electron. Another method which has recently been applied to the analysis of conventional images is that of optical diffractometry. This technique, though basically simple, is more limited in its application than stereology, since a micrograph can only be analysed if the information is contained in some sort of repeating pattern, for example a symmetrical array of identical virus particles. The principle of the method is to make an optical diffraction pattern using the transparent photographic negative of the original micrograph as the diffraction grating. The pattern is then carefully examined for all detectable forms of symmetry. Random, unrelated spots can then be treated as 'noise', and may be blanked off by inking out, or by cutting holes in an opaque mask which will only let light through those areas of the pattern which are judged to be significant. The resulting 'cleaned-up' diffraction pattern, containing only the information relating to the repeating structures in the original, is then re-formed to a conventional image by the reverse process of recombination, yielding an image of greatly increased contrast, resolution and clarity. This potentially very powerful method for resolving the structure of macromolecules down to the molecular level is described by Professor Robert Horne of the John Innes Institute in Norwich.

Both the preceding methods are applicable to conventional flood-beam microscopy. In each case, the final photographic image forms the basis of the analysis. A method whereby the direct image can be analysed before recording is quantitative microfluorometry. Modern electronics has led to the development of highly sensitive light detectors which can detect, count and classify single photons, an increase in sensitivity over the human eye of a factor of about five. These photomultiplier detectors can be placed in the image plane of a conventional light microscope, resulting in an apparatus which can directly quantitate light. Single molecules of fluorogenic substances in microscopic droplets can now be detected and quantitated. Such fluorescent compounds can in many cases be conjugated quite readily with cell components such as DNA, or with protein such as antibodies. Sections and smears of cells which have been thus treated are then excited with ultraviolet light in order to induce fluorescence in the visible region of the spectrum. The resulting fluorescent light output can then be related to the absolute amount of a known substance present in an individual cell. This technique and its applications are discussed by Dr J. S. Ploem of Leiden.

Apart from the usual classes of microscopy – light and electron, transmission and reflection – there are two basic modes of microscopy; the flood-beam mode and the scanning-beam mode. Applications of the scanning-beam mode are the major growth points of the technique of microscopy today.

The flood-beam mode is the traditional method of microscopy. A wide area of the specimen is illuminated as evenly as possible, and an enlarged image of this area is made by means of lenses. Magnification is geometrical and subject to the aberrations and imperfections of the lenses. The image is continuous in space, and is recorded photographically. The scanning mode on the other hand illuminates as small an area of the specimen as possible (the size of the illuminated area determines resolving power) with the brightest obtainable beam. The illuminating spot is moved across the width of the specimen in a sequential series of lines called a 'raster', only too familiar to viewers of 405-line black-and-white television. The image is formed sequentially point-by-point, line-by-line, as a series of electrical impulses from a detector placed close to the specimen, which can very readily be amplified and displayed on the screen of a cathode ray tube (CRT). The process is identical with that of television transmission. Image magnification is simply the ratio of the length of a single scan line on the specimen to the length of the imaging line on the display CRT. Magnifications of the order of millions are obtainable without difficulty by electronic means, entirely without the use of imaging lenses.

The scanning mode has two basic potentialities of the greatest importance to microscopy. First, the image is completely free from the aberrations and distortions to which all lenses are subject, especially electron lenses. Secondly, the image consists at one stage of a series of electrical signals, each pulse corresponding to one 'picture point'. The advantages to be gained from the first potentiality are obvious; those from the second are by no means obvious at first sight. On the contrary, it would appear to be self-evident that an image formed from a series of scan lines would be lacking in resolution compared to a sharply-focused image formed by lenses. This was indeed the case only a decade ago. Recent improvements in scanning spot formation, beam energy, and above all the application of the modern high-speed, solid-state electronic computer have so changed the position that nowadays the advantage lies almost entirely on the side of the scanning mode. The output from the photon or electron detector placed close to the specimen (above for reflection, below for transmission) varies with time as the scanning spot tracks linearly across the specimen, giving rise to a so-called 'analogue' signal. Modern electronics make it a simple matter to subdivide this analogue signal into a chosen number of discrete steps known as 'grey levels'. The signal then consists of a sequence of pulses, each corresponding to the brightness of a single picture point. This type of 'digital' signal can be processed almost instantaneously by modern high-speed computers programmed to analyse it for any number of different parameters such as size, shape, pattern, randomness, etc. The computer can be programmed to reject 'noise', to increase or reverse image contrast, to compare one set of signals with another, and to integrate, differentiate or perform other mathematical manipulations on the image signal. While the computers are performing these tasks, the digital image-forming signal is displayed simultaneously on the image CRT, giving a conventional magnified flood-beam image which the eye can interpret immediately.

The resolving power of the scanning-beam mode depends on the size of a single picture point in the image, which is a magnified image of the scanning spot. In the case of the light microscope, this is ultimately the diameter of the Airy disc imposed by the wavelength of the light used. In the case of the electron microscope, the diameter of the scanning spot depends ultimately on the spherical aberration present in the probe-forming (objective) lens, provided the source brightness can be made sufficiently high and the scanning time sufficiently long. Ultimate resolving power using the scanning mode is therefore the same as in the flood-beam mode; neither better nor worse.

Direct image analysis using the scanning mode has been applied with great success to light microscopy. Scanning microscopes based on television technology can be linked directly (or 'interfaced') with immensely sophisticated high-speed computer-processor modules, enabling image features such as particle numbers, integrated areas of a given optical density, particle shapes and sizes and particle distribution patterns to be displayed directly as numbers on a digital read-out, or punched on cards or paper tapes, or stored on magnetic tapes, or printed out on a typewriter, or plotted directly as curves on the face of a CRT; the possibilities are rapidly becoming almost limitless. The advantages in speed, accuracy and elimination of operator error to be gained by applying this automatic classification technology to routine analyses such as cell counts – performed in the past by hand and eye with an immense expenditure of time and tedium on the part of the operator – are obvious. All the basic methods of stereology can be applied to automatic image analysis, yielding stereological data far more rapidly than can point counting. The automatic image analyser has nowhere near the same powers of discrimination and pattern recognition as the human eye and brain, but in many applications its speed and accuracy are vastly superior. Dr Savile Bradbury of Oxford explains the principles and apparatus used in typical optical image analysers.

In the light microscope, a stationary spot or a mechanically-scanned 'flying spot' can be applied to the measurement of the optical density of very small objects such as cell nuclei. Many substances of great biological importance, such as haemoglobin or chlorophyll, absorb light of a specific wavelength; other substances such as DNA can be stained histochemically to do the same thing. Optical density measurements then enable the actual amount of the stained substance present in the section or smear of the cell to be calculated. Flying-spot light microscopes are now available which measure optical density and show it as a direct read-out. The addition of a simple computer enables total absorbance to be integrated, and thus the total amount of the measured substance present in any microscopical field can also be shown as a direct read-out. This technique, known as 'integrating microdensitometry', is described in the article by Dr D. J. Goldstein of Sheffield.

In a light microscope specimen, the refractive index at any point is a function of the total dry mass present in that area of the specimen. Flying-spot microscopes can therefore also be built as microinterferometers measuring refractive index, enabling the dry mass of an unstained cell component to be calculated and displayed as an integrated digital read-out. This technique, known as 'scanning microinterferometry' is dealt with in the subsequent article by Dr Goldstein.

The scanning mode, and especially the scanning transmission mode, offers even greater advantages in electron microscopy than it does in light microscopy. This arises from the fact that there are no image-forming lenses; magnification is obtained electronically and is simply the ratio of the length of the display line to the length of the scan line. A 100 nm scan line displayed 10 cm long on the face of the image CRT corresponds directly to a magnification of one million times. The absence of image-forming lenses leads to the absence of imperfections in the image due to lens aberrations. The most important of these in the conventional transmission electron microscope (TEM) is chromatic aberration. Any TEM specimen of finite thickness introduces chromatic aberration because of the reduction in velocity of those emergent electrons which have interacted with the specimen. These electrons, of course, are the ones which carry the most information about the specimen. In the normal TEM, great efforts are made to remove these electrons from the transmitted beam by using an objective aperture. In spite of this, sufficient chromatically-aberrated electrons get through to the final image to reduce the instrumental resolving power by factor of about 10; 3 nm rather than 0.3 nm. The thicker the specimen, the worse the resolution. Using the scanning transmission mode, the effect of chromatic aberration can be virtually eliminated in the simplest possible fashion. This is done by sorting the slow electrons from the fast ones by means of an electron velocity spectrometer placed after the specimen. The spectrometer enables a dark-field image at high resolution to be obtained from electrons of a given selected energy, something which cannot be done in the TEM. The resulting dark-field image is of much higher contrast than the low-resolution dark-field image obtainable in the TEM. It also enables elastically scattered electrons to be separated from inelastically scattered ones, each giving rise to a separate signal or channel. By combining these signals it is possible to obtain a signal the intensity of which is dependent on the atomic number of the atoms constituting the specimen. This multi-channel mode of operation has great potential advantages over the single-channel mode of the TEM. In addition, of course, the signal is of a digital nature, allowing direct computer interfacing with all the consequent advantages which have already been enumerated above.

STEM has an enormous potential in biology. The great increase in contrast and resolution enables thinner specimens, e.g. single molecules such as DNA, to be successfully imaged in the natural condition, free from the artifacts of staining or shadowing. When the STEM has been developed a little further, we may see an outpouring of new biological information at the molecular level comparable with the great advances in morphological ultrastructural information made with the TEM in the nineteen-fifties. The biological possibilities of STEM have so far only been touched upon, and an appraisal of its potential use in biology is given in the article by Dr H. Y. Elder of Glasgow. The theory, construction and operation of the scanning transmission electron microscope (STEM) are described in the article by Professor R. E. Burge of London.

The analysis of the energies of the electrons transmitted by a specimen is a valuable source of elemental and chemical information. The STEM instrument lends itself well to electron energy loss spectrometry, a very new and almost untried method of analysis which may well have great biological potentialities. The technique uses to advantage the scattered, chromatically-aberrated electrons which are such a headache in the TEM, but which in reality carry a great part of the information we are seeking about the chemical composition of the specimen. Energy loss spectrometry can in theory be made several orders of magnitude more sensitive than X-ray microanalysis, especially for the light elements below sodium which are so important in biology. This very new and promising technique is described by Professor R. P. Ferrier of Glasgow.

The remainder of this volume is devoted to X-ray microanalysis, a field in which remarkable progress in quantitation has been made over the past decade. X-rays are always generated as a consequence of bombarding matter with high-energy particles such as electrons. Some of these X-rays, the so-called 'Bremsstrahlungen' or 'braking rays', are caused by the generalised slowing-down of particles by matter, and form a continuous spectrum. Others, however, form peaks above this background continuum. The pattern of peaks is unique for each element in the periodic table, thus enabling it to be identified qualitatively in the specimen. The relative heights of the peaks formed by various elements are proportional to the absolute amounts of each element present, thus enabling a quantitative analysis to be made. The spatial resolution of the method depends on the volume of the specimen from which the X-rays are being excited; the projected area of this volume is considerably greater than the diameter of the exciting electron probe spot. The thinner the specimen and the smaller the spot, the higher is the spatial resolution. The method may be used with a stationary or a scanning probe spot. The former mode allows a full analysis of all detectable elements (at present, the lowest practicable atomic number is that of sodium) to be made in a volume of tissue as small as 100 nm^3. The second mode allows an image of a chosen area of the specimen to be built up in the 'light' of a single element such as calcium. X-ray microanalysis is applicable equally in the reflection and the transmission mode. X-ray microanalysers can be fitted to the specimen chambers of most commercial TEMs and SEMs. There are two main types of X-ray detector now in use; the energy-dispersive (ED) and the wave-length dispersive (WD) systems. The former uses a semi-conductor detector which gives an electrical signal proportional to the energy of each impinging X-ray photon, energy being inversely proportional to wavelength. The composite signal emerging from the ED detector contains simultaneous signals from the whole detectable wavelength spectrum (peaks plus continuum); this is displayed as a histogram on a CRT, thus enabling a complete qualitative and quantitative analysis of the elemental composition of the specimen to be built up in a matter of seconds. Although the sensitivity and convenience of the ED system are much better than those of the more complex and time-consuming WD system, peak-to-peak resolution is poorer. Each has its own advantages and disadvantages. In this volume, the ED system is discussed by Professor T. Hutchinson of Minnesota, and the WD system by Dr J. Chandler of Cardiff.

If one is to obtain any meaningful results by X-ray or electron spectrometry concerning the distribution of elements and compounds in a section of biological material, it is essential that their distribution in the section should correspond accurately with the distribution in the living material. This applies particularly to soluble compounds such as salts of sodium and potassium for example. This rules out the use of all dehydrated or embedded material, as the processes of fixation, embedding and staining generally remove soluble compounds and radically alter the distribution of others, at the same time adding substances which were not originally present. One approach to a solution of this problem is to cut frozen ultrathin sections of material, preferably unfixed, and to freeze-dry them before using the sections as specimens for microanalysis, or even to use them in the hydrated state. The techniques for preparing ultrathin sections of frozen material are known as 'cryo-ultramicrotomy' and are described in the concluding chapter by Dr T. C. Appleton of Cambridge.

N. T. JAMES

Stereology

Stereology can be defined as the science whereby three-dimensional information on structures can be derived from only two-dimensional sections of those structures. Stereology can be used in the study of structures independently of their size or their composition. Stereological techniques can be used to study problems ranging from the distribution of stars in space to estimating the proportions of soil and roots beneath trees. The same formulae and techniques of stereology can be used to study the relative proportions of different rocks in geological specimens and also the changes in the content of intracellular organelles which can be induced by experimental procedures or pathological processes.

Biology is one of the least numerate of the sciences. Quantitation in general, and particularly that provided by stereology, is capable of providing absolute values so that we know much more about the structures that interest us. It enables different specimens to be compared to see whether real differences exist between them. Stereology also has the advantage that it enables us to find changes in cell populations and tissues that would otherwise escape detection. Though workers in the materials sciences have tended to dominate the development of stereology, in the last decade it has become one of the most powerful analytical techniques in biology. This is particularly evident in the discipline of cell biology. There are numerous textbooks and reviews which amply illustrate the power and applicability of stereology (DeHoff & Rhines, 1968; Underwood, 1968, 1970; Elias, Hennig & Schwartz, 1971; Weibel & Bolender, 1973).

In practice, stereological analyses proceed in several distinct stages. Adequate representative sections of the tissue are first obtained. Test lattices are then superimposed upon photographs, micrographs or projected images of these sections. Each test lattice consists of a transparent plastic sheet bearing a line or point pattern. The final stage of analysis consists of recording the interaction between the test lattice and the photomicrographs of the test sections. This may mean making counts of the proportion of points in a test lattice which fall upon a feature of interest or counting the number of intersections that the surface profiles of a cell organelle make with a test lattice consisting of straight lines. The type of test lattice used will depend on the particular stereological parameter being estimated. Examples will be given later.

As an alternative procedure to using test lattices formed from transparent plastic sheets, complex electronic image-analysing devices, for example the Quantimet (see Bradbury, this volume) could be used. All such devices, however, are based firmly upon stereological principles for both their design and use. The direct application of test lattices to photographs or photomicrographs is relatively cheap and enables any research worker to begin collecting stereological data without using expensive apparatus. Once the biological problem has been defined, the specimens prepared and the appropriate test lattices drawn, then data collecting can begin. Image analysers, however, have the advantage that many point or line intersection counts can be made in a short time so that much more tissue can be examined. A list of the stereological parameters commonly used to quantify cells and tissues is given in Table 1.

Many structural parameters of cells and tissues can be quantified using stereological techniques. The following brief selection gives some typical examples from current biological research

(*a*) The sizes of nuclei, their volume fraction (V_V) in various cells and tissues and their number (N_V) per unit test volume of tissue.
(*b*) The size–frequency distributions of small particles, e.g. neurotransmitter vesicles within nerve axons.
(*c*) Surface areas of cell membranes (S_V) per unit test volume of tissue, e.g. for mitochondrial cristae.
(*d*) The total length of capillaries (L_V) per unit test volume of tissue.
(*e*) The degree of orientation (Ω) of particles, e.g. the orientated mitochondria in skeletal muscle fibres.

In order to illustrate the concepts and techniques of stereology only a few of the more commonly used stereological parameters will be considered here, e.g. estimates of V_V, S_V, L_V and N_V, although several other basic problems in stereology will also be considered. A frequently occurring problem is that of estimating the size–frequency distribution of particle populations. With particles of varying sizes the determination of true particle size from tissue sections can be difficult. The problem becomes particularly difficult when the average size of particles and section thickness are of the same order. However, several correction formulae are available for determining true sizes more accurately. Recently it has become clear that the distribution of organelles within cells may be of some importance. Consequently some aspects of pattern analysis will be illustrated.

Table 1. *Definition of terms used in stereology*

Symbol	Dimensions (arbitrarily given in μm)	Definition
P	—	Number of test points
P_P	—	Point fraction. Number of points on areal feature of interest per test point
P_L	μm^{-1}	Number of point intersections per unit length of test line
P_A	μm^{-2}	Number of points per unit test area
P_V	μm^{-3}	Number of points per unit test volume
L	μm	Test line length
L_L	$\mu m\ \mu m^{-1}$	Lineal fraction. Length of lineal intercepts per unit length of test line
L_A	$\mu m\ \mu m^{-2}$	Length of lineal elements per unit test area
L_V	$\mu m\ \mu m^{-3}$	Length of lineal elements per unit test volume
A	μm^2	Test area
S	μm^2	Surface area
A_A	$\mu m^2\ \mu m^2$	Areal fraction. Area of feature of interest per unit test area
S_V	$\mu m^2\ \mu m^{-3}$	Surface area per unit test volume
V	μm^3	Volume
V_V	$\mu m^3\ \mu m^{-3}$	Volume fraction. Volume of feature of interest per unit test volume
N	—	Number of features of interest
N_L	μm^{-1}	Number of intersections of features per unit length of test line
N_A	μm^{-2}	Number of intersections of features per unit test area
N_V	μm^{-3}	Number of features per unit test volume
L_3	μm	Average lineal intercept, L_L/N_L, across feature of interest

An intersection is defined as occurring where a profile is crossed by a test line of a stereological lattice.

An intercept is defined as the length of a randomly distributed straight line that lies within the boundary of a particle.

Stereological principles

Dimensional relationships

When a three-dimensional cell or tissue is sectioned very thinly and examined with a microscope then only two-dimensional profiles are seen (Fig. 1). The two-dimensional structures present become observable as only one-dimensional structures. One-dimensional structures are reduced to points of zero dimensions. As a rule the occurrence of an n-dimensional structure is seen in a section as an $(n-1)$-dimensional structure. With this simple rule of

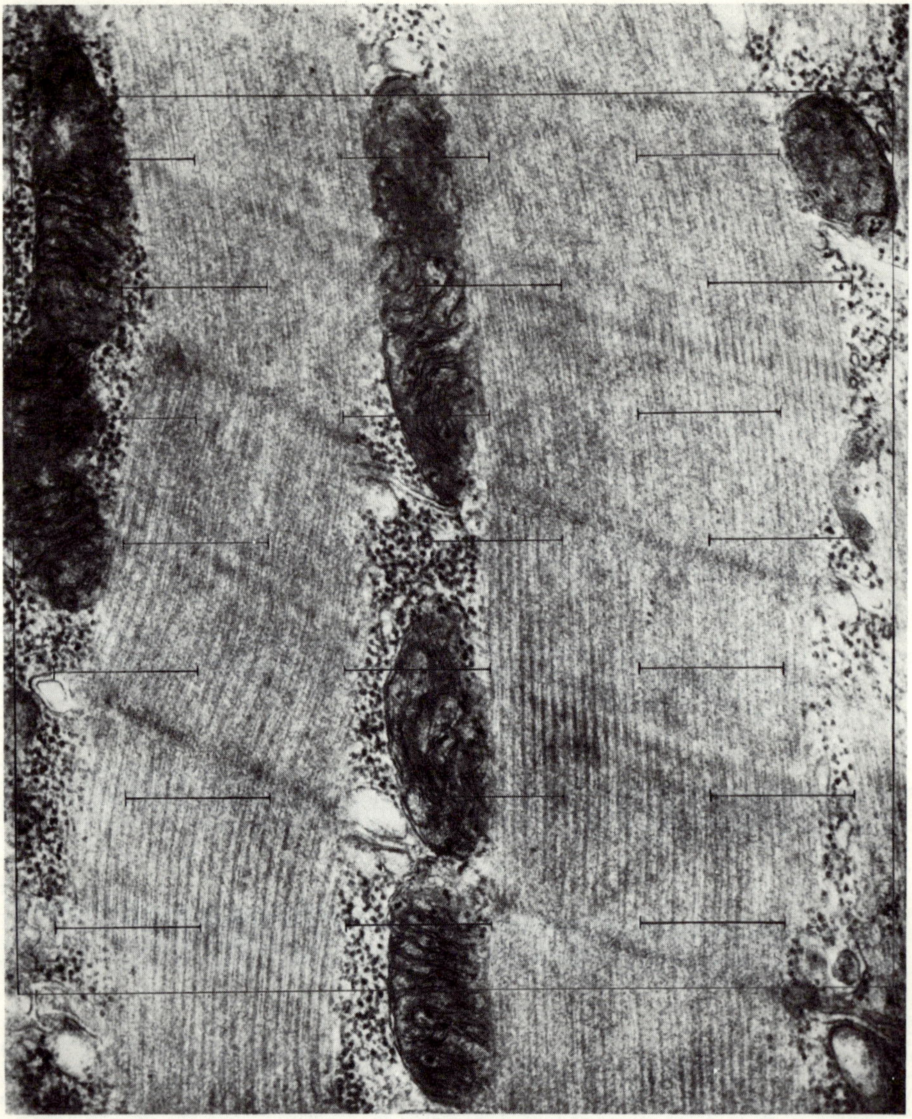

Fig. 1. Electron micrograph of a longitudinal section of skeletal muscle fibre with superimposed test lattice. The test lattice bears twenty-one short test lines each equivalent to $\sim$ 0.57 μm of tissue. They lie perpendicular to the long axis of the muscle fibre. The area of the test lattice is equivalent to $\sim$ 12.2 μm^2. The end of each test line can be regarded as a point. Forty-two points lie in the test area and each is equivalent to an area of $\sim$ 0.29 μm^2. Five points lie on mitochondria and there are thirteen intersections of test line with mitochondrial profiles. The areal fraction of the test lattice

dimensional reduction most structures can be accurately identified in sections. For example, the three-dimensional volume occupied by a nucleus becomes visible as a two-dimensional area, and a randomly sectioned two-dimensional structure appears in one dimension as a line.

The numerical reconstruction of cell structure by stereological analyses proceeds exactly contrary to the rule of dimensional reduction. Data collected in one dimension are used as an exact predictor of two-dimensional properties of structures. Two-dimensional data, for example areal counts of nuclear profiles, are used to provide exact three-dimensional data on their volume fractions.

Table 2 illustrates the relationship between experimentally measured properties, for example (P_P), (L_L) and (L_A), to the calculated higher dimensional parameters, for example $\boxed{V_V}$, $\boxed{S_V}$ and $\boxed{L_V}$.

Practical measurements

Measurement of P_P. A test lattice composed of a regular pattern of points arranged in square array is used for this measurement. A test lattice consisting of equally spaced horizontal and vertical lines can also be used. In this case, the line intersections can be used as points. The test lattice must contain a finite boundary and a known number of points. The test lattice boundary must be enclosed within the boundary of the test tissue during use. Consider a structure of interest, e.g. the nucleus. The test lattice is randomly placed on a micrograph of the test section and the number of points falling on nuclei are recorded. If desired, the test lattice can then be removed and replaced in a new random position overlying the micrograph and a fresh point count made. The total number of points P_α falling on nuclei during the test count when divided by the known number of test lattice points (P_T) gives the point fraction ratio (P_P) for the test nuclei. For example if $P_T = 200$ and $P_\alpha = 31$ then $P_P = 0.16$. If the test lattice consists of points arranged in a square array, then provided the distance between the lattice points and the magnification of the micrograph are known, absolute sizes of structures can subsequently be calculated (see later).

---

occupied by mitochondria is 0.12 (5/42) and $(P_L)_\perp$ is 1.09 μm^{-1} ($13/21 \times 1/0.57$). If the tissue were isotropic then V_V would equal 0.12 and S_V would equal 2.18 $\mu m^2\ \mu m^{-3}$ ($2\,(P_L)_\perp$). However, since muscle fibres are anisotropic, if the lattice is rotated through 90° and the number of intersections of test line with mitochondrial profiles is counted to give $(P_L)_\parallel$, then S_V is given by the term

$$\frac{\pi}{2}(P_L)_\perp + 2(P_L)_\parallel - \frac{\pi}{2}(P_L)_\parallel\ \mu m^2\ \mu m^{-3}.$$

Table 2. *Relationship between experimentally measured properties* (○) *and calculated parameters of tissues* (□)

Microscopical features	Dimensions of symbols (arbitrarily expressed in μm) μm^0	μm^{-1}	μm^{-2}	μm^{-3}
Points	P_P	P_L	P_A	P_V
Lines	L_L	L_A	L_V	—
Surfaces	A_A	S_V	—	—
Volumes	V_V	—	—	—

The equations relating the quantities in Table 2 can be expressed as follows:

In the first column
$V_V = A_A = L_L = P_P \quad \mu m^0$.
In the second column

$$S_V = \left(\frac{4}{\pi}\right) . L_A \ \mu m^{-1}$$

and

$$L_A = \left(\frac{\pi}{2}\right) P_L \ \mu m^{-1}.$$

The quantities listed in these two columns are connected by the powerful relationship

$$S_V = \frac{2}{L_3}.$$

where L_3 is defined as the mean intercept length across test particles, which is given by

$$L_3 = \frac{1}{N_L}.$$

In the third column
$L_V = 2P_A \ \mu m^{-2}$.

The rows of quantities lying in the second, third and fourth columns are connected by the expression

$$P_V = 2P_A P_L \ \mu m^{-3}$$

for the horizontal row, and by

$$P_V = \left(\frac{L_V S_V}{2}\right) \mu m^{-3}$$

for the oblique row.

Measurement of P_L. An important measurement frequently used in stereological analyses is the number of intersections of a surface per unit length of test line in a test lattice. The test lattice possesses a finite boundary within which a known total length of test lines has been drawn. The test lattice boundary is completely enclosed within the boundary of the test tissue. The test lattice is randomly placed on a micrograph of the test tissue and the number of times that, for example, the surfaces of nuclei intersect the test lines, is recorded. As an example, suppose that on a micrograph of magnification 10^4, the number of intersections made by the test lines with nuclear surfaces averages 32 for a test line length of 10 cm. Since the test line length represents 10 μm of the tissue, then $P_L = 3.2\ \mu m^{-1}$.

Derivation of areal fractions (A_A) from point fractions (P_P). If a test point is placed at random within a test area A_T containing a feature of interest of area A_α, then the probability of the point falling within the feature boundary (P) is equal to the ratio of the two areas. Consequently

$$P = \frac{A_\alpha}{A_T}.$$

For many randomly placed test points P_T the expected value $E(P_\alpha)$ of the number of points lying within the feature boundary is given by

$$E(P_\alpha) = P_T \cdot P = P_T\left(\frac{A_\alpha}{A_T}\right).$$

Therefore

$$\frac{A_\alpha}{A_T} = \frac{E(P_\alpha)}{P_T}$$

or

$$A_A = P_P.$$

When the test points are distributed in a regular pattern the analysis is termed a systematic point count. A systematic two-dimensional point count is the most efficient method of volume fraction analysis. Random point counts yield a larger coefficient of variation in the final result when compared with a systematic point count at equivalent test lattice density and for the same total number of points used. For maximum efficiency the number of occupied points should be approximately equal to the number of features analysed under the test lattice.

Derivation of volume fraction (V_V) from areal fraction (A_A). The volume fraction (V_V) is one of the most important and commonly used parameters for characterising tissues and cells. The French geologist Delesse (1848) was the first to indicate that the areal fraction of a component lying in a random transverse section taken through a rock is equivalent to the volume fraction of that component throughout the rock. To quote Delesse: 'Let us assume that the volume occupied by the rock is related to a system of co-ordinates, and let P be the area occupied by one of the constituent minerals in a section formed by a plane parallel to *xy*. To obtain the exact volume occupied by this mineral in the rock, it would be necessary to know the successive values of P for a series of infinitely close planes parallel to *xy*. The integral $\int P \cdot dz$ would then give the volume sought.'

Herein lies the exact principle for establishing that

$$A_A = V_V.$$

This relationship, as with many others in stereology, has been independently derived on many occasions. Knowledge of its derivation and also the derivations of other relations are of considerable value in understanding clearly the general principles of stereology.

Consider a test cube of volume V_T that possesses a side of length l. Clearly,

$$V_T = l^3.$$

Let the test cube contain a number of randomly distributed and irregularly shaped structures denoted by α. Suppose a thin transverse section of thickness δx be cut transversely across the cube. The area of the section will be l^2 and its volume will be $l^2 \cdot \delta x$. The volume of the structure is given by

$$\delta V = l^2 \, \delta x (V_V)_\alpha \ \mu m^3.$$

Where δx becomes infinitely thin

$$\delta V_\alpha = A_\alpha(x) \, \delta x \ \mu m^3,$$

where $A(x)$ varies with the position (x) of the section. Its average value between one edge of the test cube and the other at a distance l is given by

$$\bar{A}_\alpha = \frac{1}{l} \int_0^l A_\alpha(x) \, dx \ \mu m^2.$$

The total volume of the structure, V_α, is given by

$$V_\alpha = \int_0^l A(x) \, dx,$$

which becomes

$$V_\alpha = l\bar{A}_\alpha.$$

Dividing both sides by V_T we obtain

$$\frac{V_\alpha}{V_T} = \frac{\bar{A}_\alpha}{A_T}$$

or, in general,

$$V_V = A_A.$$

This relationship indicates that the average areal fraction represents a close estimate of the volume fraction of the analysed structure. Although one co-ordinate was used to specify the location of the section plane in the derivation, the final relationship is independent of any such restriction.

It should be noted that since

$$P_P = A_A$$

then V_V can be estimated directly by the relation

$$P_P = V_V,$$

where P_P is measured directly from micrographs.

Derivations of the other stereological relationships are readily available in the literature (DeHoff & Rhines, 1968; Underwood, 1970).

Volume fraction (V_V) estimations in anisotropic tissues. In many cell types it is clear that their organelles have a specific non-random distribution. For example, in many skeletal muscle fibres the mitochondria tend to predominate in I-band regions rather than in A-band regions and their distribution confers a longitudinal anisotropy on the fibres. The mitochondria are also distributed unevenly across the transverse plane of many fibres. In some fibres the mitochondria seem to occur just below the surface whilst in others they occur as a single central core in the long axis of the fibre. This type of distribution confers a transverse anisotropy on muscle fibres. In order to compensate for the longitudinal anisotropy a large number of randomly cut transverse sections would be required so that the relative lengths of the A-bands and I-bands would be represented in the final sample. As an alternative procedure a smaller number of oblique sections extending over a dozen or so sarcomeres are sufficient. In order to overcome the transverse anisotropy the area of the fibre sampled must be representative of the entire cross-sectional area of the fibre. Either a large number of random sample areas or a symmetrical estimator of the cross-section of the fibre must be obtained. The only accurate estimator available is a sector taken from the cross-section of the fibre. In summary, for minimum effort skeletal muscle V_V estimations can be made on randomly orientated oblique sectors of muscle fibre sections rather than on numerous randomly selected fields taken from an equally large number of random oblique sections. Theoretically, oblique sections should not interfere with mitochondrial V_V estimates in skeletal muscle (Weibel, 1972). Other less time-consuming methods of sampling will probably be derived for other anisotropic tissues.

Surface area (S_V) estimates. Some typical surfaces of biological interest which can be analysed using stereological techniques are those of nuclei, mitochondria and also those of small particles, for example neurotransmitter vesicles or secretion droplets.

The basic equation for obtaining the surface areas of such structures in a test volume is given by

$$S_V = 2P_L \mu m^2 \mu m^{-3}.$$

This equation applies to surfaces with any configuration and is valid for surfaces bounded by other features and for continuous networks of surfaces. An alternative formula using the number of organelle intersections per unit length of test line (N_L) is given by

$$S_V = 4N_L \mu m^2 \mu m^{-3}.$$

The formulae are valid for both convex and concave structures though where a test line is re-entrant (i.e. crosses a structure boundary more than once) only one organelle intersect is counted.

In order to evaluate S_V, only a simple intersection count (P_L) with organelle profiles in a test section is required. Complete randomness between the intersecting test lattice and the surfaces is required.

Surface area (S_V) estimates in anisotropic tissues. If the tissue is markedly anisotropic and there is any physical relation between the plane section and the axis of anisotropy then the formula $S_V = 2P_L$ is not valid. An alternative formula must be derived which is specific for counts made when the test lattice lies in the orientation plane of an isometric tissue. A typical example occurs when analysing longitudinal sections of skeletal muscle. Several other tissues also seem highly orientated, for example many types of neurone.

Theoretically, there are three types of orientated structure. Two of these types, the orientation of lines in a plane and the orientation of lines in space, are of relatively little interest to biologists. The most important type of orientation, that of surfaces in space, is more relevant to biological problems. Surfaces play an important role in biological activities, for example they play a role in diffusion processes and may carry numerous enzymes. Frequently the extent of their surfaces must be known quantitatively. Surfaces can either be fully orientated, a condition which rarely occurs in living tissue, or partially orientated. For example, the surfaces of nuclei and mitochondria are partially orientated with respect to the long axis of muscle fibres.

If only the total surface area of an orientated system of surfaces were required, then provided a sufficient number of random sections were available the relationship $S_V = 2P_L \mu m^2 \mu m^{-3}$ holds. However, we may also be interested in the degree or type of orientation of the surfaces. In addition, there may be other reasons why it is more convenient or simpler to cut sections in only one plane. For example, there may only be small amounts of tissue available for analysis. Such is the case for the small and infrequently occurring sensory organs found in skeletal muscle, the muscle spindles. Expressions for S_V have been derived which use data derived from sections cut in a single plane lying in the long axis of orientated structures. The equation for S_V of a partially orientated surface with a linear orientation is given by

$$S_V = \frac{\pi}{2}(P_L)_\perp + 2(P_L)_\parallel - \frac{\pi}{2}(P_L)_\parallel \ \mu m^2 \ \mu m^{-3}.$$

A test lattice consisting of a number of short straight lines is used to estimate S_V. The lattice is superimposed on test sections and the number of intersections of surface profiles with the test lattice is recorded. If the number of intersections is independent of the orientation of the lattice then the surface profiles are randomly orientated and the surfaces are termed *isometric*. For an orientated system of surfaces the number of intersections will vary markedly with the orientation of the test lattice. Such a system of surfaces is termed *anisometric*.

Only one section plane is required for the analysis of partially orientated surfaces with a linear orientation, i.e. a longitudinal section parallel with axis of orientation. On test sections a linear test lattice is applied in two directions, first parallel and then perpendicular to the orientation axis. In the first application, the orientated surfaces tend to lie parallel with the orientation axis and therefore they also tend to lie parallel with the lines of the test lattice. Consequently, only the isometric portion of the surfaces will be intersected and we may write

$$(S_V)_{iso} = 2(P_L)_\parallel \ \mu m^2 \ \mu m^{-3},$$

where $(S_V)_{iso}$ is the surface area of the isometric surfaces per unit volume of test tissue and $(P_L)_\parallel$ is the number of intersections per unit length of test line that the surfaces make with the test lattice. In the second application, the test lattice is reorientated so that the test lines lie perpendicular to the axis of orientation. The number of intersections counted in this direction, $(P_L)_\perp$, reflects both the isometric and the linearly orientated surfaces. Consequently, the area of the linearly orientated surfaces, $(S_V)_{lin}$, is given by

$$(S_V)_{lin} = [(P_L)_\perp - (P_L)_\parallel] \ \mu m^2 \ \mu m^{-3}.$$

The total surface area per unit volume of test tissue is given by

$$S_V = (S_V)_{iso} + (S_V)_{lin},$$

which is equivalent to

$$S_V = \frac{\pi}{2}(P_L)_\perp + 2(P_L)_\parallel + \frac{\pi}{2}(P_L)_\parallel \ \mu m^2 \ \mu m^{-3}.$$

Note that $(P_L)_\perp$ and $(P_L)_\parallel$ are counts made on the same test fields in a single longitudinal plane of section. The degree of orientation of the surfaces, Ω, can be defined as

$$\Omega = \frac{(S_V)_{lin}}{(S_V)_{lin}+(S_V)_{iso}}.$$

Using the same values for $(P_L)_{\perp}$ and $(P_L)_{\parallel}$ we obtain

$$\Omega = \frac{(P_L)_{\perp}+(P_L)_{\parallel}}{(P_L)_{\perp}+\frac{4}{\pi}(P_L)_{\parallel}-(P_L)_{\parallel}}.$$

Sampling techniques

Experimental design

An absolute requirement of stereology is that the test samples should be representative of the test tissue. Test samples should therefore be taken completely randomly both from the experimental animals and from the test tissues. Since this ideal procedure is often difficult or impractical to follow in practice, a systematic random sampling technique is generally preferable.

As in all biological work the design of the experimental or sampling programme should be considered carefully in order to obtain greatest accuracy. It should also be possible to provide some estimate of the financial cost and work load involved. Closely related to the sampling programme is the task of estimating the number of points necessary for the stereological analysis. Consider the following theoretical example of a systematic random sampling programme.

Sampling stage	Specimen totals
n_1...6 animals	6
n_2...6 tissue blocks from each animal	$6\times6 = 36$
n_3...6 sections from each block	$6\times6\times6 = 216$
n_4...6 micrographs from each section	$6\times6\times6\times6 = 1296$

The choice of six specimens at each stage is purely arbitrary and may be varied according to the choice of the investigator. Some estimate of the effect of varying the relative numbers in each sampling stage can readily be obtained. Let the variance of the mean value obtained by analysing 1296 micrographs be $V(\bar{x})$. Let the variances between n_1 animals, n_2 tissue blocks, n_3 sections and n_4 micrographs be S_1^2, S_2^2, S_3^2, and S_4^2 respectively, where S is the standard deviation of each population. We find that

$$V(x) = \frac{S_1^2}{n_1}+\frac{S_2^2}{n_1 n_2}+\frac{S_3^2}{n_1 n_2 n_3}+\frac{S_4^2}{n_1 n_2 n_3 n_4}.$$

It can be seen that though increases in any value of n reduce the total variance $V(x)$, increasing n_1, n_2, n_3 and n_4 respectively has successively less effect on $V(x)$. An additional variance (S_5^2) relating to the number of test points per field (n_5) may also be added to the value for $V(x)$. This final term, given by

$$\frac{S_5^2}{n_1 n_2 n_3 n_4 n_5}$$

indicates the relatively insignificant effect of increasing the number of test points applied to the micrographs compared with increasing n_1, n_2, n_3 or n_4 (Weibel, Kistler & Scherle, 1966).

In summary, if the sections or electron micrographs are atypical due to poor sampling technique, then increasing the number of test points cannot compensate. Some consideration must now be given to the number of test points which need to be used to achieve the desired accuracy with the minimum effort.

Minimum number of test points required in stereological analyses

The investigator must decide at the beginning of the analysis the maximum error to be tolerated. The effort of analysis will vary inversely with the square of the error. For example, the difference between an accuracy of 10 % and 1 % could correspond to an effort of 30 minutes and 1 week respectively. In biological studies an accuracy of 10 % is often adequate.

The following general procedure is recommended. An approximate value of the volume fraction can be obtained by applying a test lattice a few times to a few randomly selected micrographs and measuring the point fraction P_P. Using this approximation the minimal number of test points, P_T, needed to achieve the desired accuracy can be calculated. For an accuracy with 95 % confidence limits DeHoff (in Underwood, 1970) suggests that

$$P_T = \frac{200\sigma\,(P_P)}{(\%\ \text{accuracy})\cdot P_P},$$

where σ is the standard deviation, whilst Gladman & Goodman (in Underwood, 1970) suggest that

$$P_T = P_P\,(1-P_P)/\sigma^2(P_P).$$

A more convenient formula is that derived by Hally (1964) and subsequently re-derived by other workers. Hally suggests that the relative standard error (RSE) of a point count can be given by the relation

$$\text{RSE} = \sqrt{\left(\frac{1-P_e}{n}\right)},$$

where P_e is the approximate estimate of P_P. The relative standard error is defined as the standard error of the point count divided by the relative area of the test component. Values for n (the minimum number of points required so that the arbitrary value for the RSE is not exceeded) can then be calculated. For example, for an RSE of 1 %, if P_e is ~ 0.35 then ~ 6500 test points are required, whereas if P_e is ~ 0.02 then the number of test points required is ~ 9800. However, if an RSE of 5 % (0.05) is acceptable, then if P_e is ~ 0.35 the number of test points required is ~ 260 whereas if P_e is ~ 0.02 the number of test points required rises to only ~ 400. The test lattice is selected so that an average of one test object per point is present.

Biased sampling

The selection of random samples from homogeneous tissues, for example the liver, is relatively easy. When more complex organs are studied, such as kidney or stomach, then random tissue samples must be taken from equivalent regions, for example the renal cortex or the muscularis mucosae. Such tissue samples still have to be randomly selected. Selection of fields for photography in the electron microscope is subject to some bias. The temptation to photograph only the aesthetic regions of a section must be avoided. A convenient method of systematic sampling is to photograph a constant region of each grid square, for example, the most left-hand corner of each grid. Though this procedure is theoretically unsound due to possible repeating structures within the tissue, further random selection of these micrographs avoids bias.

Biased component sampling

Biased samples of electron micrographs can be produced deliberately or inadvertently either by the sectioning technique or by the non-random selection of fields for microscopy. It may be necessary to cut sufficiently deeply into cells in order to recognise a particular cellular component or to identify a cell type unambiguously. Such is the case for macrophages, some neurones and cells of the lymphocyte/plasma cell family. Under these conditions the sample does not consist of random sections passing with equal probability through all regions of the cell and is termed a *component biased sample*. Stereological correction formulae are required to yield the true results using the experimentally observed but biased data.

Consider the case of a spherical cell containing a single, centrally placed spherical nucleus. If random sections are cut so that a portion of the nucleus is always included in each section then the true volume proportion $(V_V)_t$

can be derived from the observed volume proportion $(V_V)_o$ by the relation

$$(V_V)_t = \left(\frac{3(V_V)_o}{(V_V)_o+2}\right)^{\frac{3}{2}}.$$

If the nucleus is eccentrically placed within the cell then the true volume proportion is given by the relation

$$(V_V)_t = \left\{\frac{(V_V)_o + \sqrt{[33(V_V)_o^2 + 48(V_V)_o]^3}}{4(V_V)_o+6}\right\}.$$

Further details are given by Mayhew & Cruz (1973). Other component biased samples require the derivation of new stereological correction formulae.

Further stereological analyses

Measurement of the numbers of particles per unit test volume (N_V)

We often need to know the number of small granules present per unit test volume of tissue (N_V). Such problems are typical in studies on secretion granules in secretory cells or neurotransmitter vesicles in axons. The exact stereological formula relating N_V to the measured number of granules per unit test area in two-dimensional sections is given by

$$N_V = N_A/\bar{D},$$

where $\bar{D}$ is the true mean granule diameter. N_A is measured directly from the test sections.

It should be noted that the quantity measured in two dimensions (N_A) is related to the product of two independent three-dimensional quantities, viz. N_V and $\bar{D}$. The evaluation of either of these parameters requires knowledge of the other. In order to estimate N_V an independent measurement of $\bar{D}$ is needed.

There are difficulties in estimating $\bar{D}$ from two-dimensional sections. Consider a population of spherical granules. Random sectioning of the population, even when this is comprised of identically sized spheres, yields a population of profile diameters of which the mean (d) is smaller than the true mean diameter ($\bar{D}$). This arises because the granules are usually sectioned some distance from their maximum diameters. The true mean diameter can be derived from the profile mean diameter by the relationship

$$\bar{D} = \frac{\pi}{4}d.$$

If the population consists of granules of varying sizes then the estimation of $\bar{D}$ becomes more difficult. Two approximation techniques are available for determining $\bar{D}$.

In the first technique, the profile diameters are segregated into a number of classes to form a size–frequency histogram. It will be apparent that the largest class of profiles almost represents the equatorial profiles of the largest particle class. Using a size–frequenecy histogram of the number of expected profiles for a population of identically sized spheres, the contribution of this class of profiles to the smaller profile classes can be calculated and subtracted from the values in the histogram. The procedure is then progressively repeated for each smaller subsequent histogram class. Numerous methods together with the practical details and their theoretical derivations can be found in the literature (Underwood, 1970).

The second of the two techniques requires only that the size–frequency distribution of the profiles be known. Hennig & Elias (1971) have presented sample graphs of various size–frequency distributions for profiles that can be compared with experimentally observed distributions. Values for $\bar{D}$ can thus be readily obtained on inspection without complicated numerical analyses.

Correction formulae necessary when granule sizes and section thickness are comparable

When examining test granules by electron microscopy, if their diameters are large (for example, more than 1 μm) when compared with a section thickness of $\sim$ 50 nm, then the sections can be regarded as two-dimensional. However, when the granules are small compared with the section thickness, then the electron micrographs represent three-dimensional projections of the unsectioned granules within the section, and considerable overestimation of parameters such as V_V can occur as a result. The overestimation occurs because particles which lie entirely within the section are included in the micrograph, yet they do not lie in the two-dimensional surface of the section which is the only region which should be analysed. This phenomenon is often referred to as the Holmes effect.

For the volume fraction estimations of small spherical granules the true volume fraction $(V_V)_t$ can be estimated from the observed volume fraction $(V_V)_o$ by the relationship

$$(V_V)_t = (V_V)_o\left(\frac{2\bar{D}}{2\bar{D}+3t}\right),$$

where $\bar{D}$ is the mean particle diameter and t the section thickness. When the particle diameters are fifteen times the section thickness the overestimation of V_V is $\sim$ 15 %. The overestimation increases rapidly as particle diameters approach section thickness. When particle diameters are twice the section thickness the overestimation is $\sim$ 100 %.

Where the numerical density (N_V) of small granules is high, as in neurotransmitter vesicles in terminal axons, other effects such as the overlapping of projected images of the granules can also occur. The appropriate correction formula, which allows for both the Holmes effect and the overlapping of granules, is given by

$$(V_V)_t = (V_V)_o \left(\frac{\bar{D}}{\bar{D}+3t}\right).$$

Pattern analysis

Pattern analysis is closely related to the conventional type of stereological analysis. Currently, the technique of studying the distribution of test objects appears to have two basic aims. First, for limited classes of test objects it may be possible to determine whether they are randomly or non-randomly distributed. Secondly, pattern analyses may be useful in discriminating between populations of test objects where the conventional parameters of stereology fail to do so. Such studies have the potential to reveal much information on the way in which organelles are distributed within cells. They might also demonstrate how their distributions change following physiological activity or experimental procedures.

One procedure in studying the random distribution of test objects is to compare their distributional properties found by experiment with those properties which would be expected if the test objects were scattered randomly. Such a technique is used in nearest neighbour analyses. Some care should be taken over the use of the term 'random'. Random populations of test objects have precisely defined properties. For example, if we consider the random distribution of points in two and three dimensions we find that the mean distance of each point to its nearest neighbour ($\bar{x}$) is given by the term

$$\bar{x} = \frac{1}{2\cdot\sqrt{(N_A)}} \pm \text{S.E.}\sqrt{\left(\frac{4-\pi}{4\pi n N_A}\right)},$$

where n is the total number of points in the population. For points scattered in three dimensions $\bar{x}$ is given by the term

$$\bar{x} = \frac{0.554}{3\sqrt{(N_V)}}.$$

For any population of points in two dimensions, if the mean nearest neighbour distance found by experiment differs significantly from the calculated value then the population is clearly non-randomly distributed. Some 'pattern' is present even though it is not visible on inspection. The technique has been used to show that longitudinal columns of mitochondria in skeletal muscle are not randomly distributed in the transverse plane of the fibres. Rarely, however, are two-dimensional distributions of test objects found in biological tissue other than in markedly anisotropic tissue. The finding of non-random distributions naturally provokes thought as to the reasons for their occurrence.

Unfortunately there are two major disadvantages with nearest neighbour analyses. First, they are applicable only to populations of test objects that can be approximated to populations of points. The distributional properties of objects which constitute a large volume fraction of the test structure are unknown. The second objection is that test objects are distributed usually in three-dimensional space. The true nearest neighbour distance is rarely included on two-dimensional test sections. The apparent nearest neighbour distances measured on test sections are considerable overestimates of the true values. Whilst nearest neighbour analyses are of relatively limited practical use at present, they clearly illustrate the necessary properties in a test population in order for it to be *concluded* from experimentally derived data that the population is either randomly or non-randomly distributed.

A more convenient parameter, though little used, is the mean free distance between particles (λ). The mean free distance is defined by the term

$$\lambda = \frac{1-V_V}{N_L}\,\mu\text{m},$$

where V_V and N_L are the usual parameters. This relation follows from the fact that the number of intersections with the test objects (N_L) is the same as the number of intersects with the surrounding matrix. The lineal fraction occupied by the matrix is N_L which is also equal to its volume fraction $(1-V_V)$. The mean free distance is an edge-to-edge spacing and represents the uninterrupted interparticle distance through the matrix averaged between all the possible pairs of particles. The relation is valid regardless of size, shape and distribution of the test objects.

A further stereological relation useful for pattern analysis is given by

$$\lambda = L_3\left(\frac{1-V_V}{V_V}\right).$$

This relation could be exploited further in pattern analysis since with constant V_V either λ or L_3 could be used as a distribution index for any test population. Two populations of cells possessing similar values for V_A of mitochondria might differ considerably in the distributional pattern of these organelles and have markedly different values for λ.

Cruz (1976) has derived a new approach to the pattern analysis of test objects. His paper should be consulted for further details. Instead of test lattices consisting of points or lines, lattices are constructed which consist of circles arranged in regular patterns. For convenience, a series of lattices is prepared each bearing circles of a different size. These lattices are then superimposed on micrographs as with conventional stereological analyses.

Consider two different test object populations arbitrarily termed X and Y which lie within a given cell type. For each application of the test lattice the number of circles that intersect with both X and Y separately are recorded. In addition the number of circles which intersect X and Y simultaneously are also recorded. The following parameters are then calculated:

$$a_X = \frac{\text{Total number of circles intersecting X but not Y}}{\text{Total number of circles intersecting X and/or Y}},$$

$$a_Y = \frac{\text{Total number of circles intersecting Y but not X}}{\text{Total number of circles intersecting X and/or Y}},$$

$$a_{XY} = \frac{\text{Total number of circles intersecting X and Y}}{\text{Total number of circles intersecting X and/or Y}}.$$

Two test specimens, for example, two cell populations termed G and H, can then be analysed as follows:

The parameter T is calculated using the relationship

$$T = \frac{(a_G - a_H)}{\sqrt{\left[\frac{a_G^2(1+a_G)}{N_G} + \frac{a_H^2(1+a_H)}{N_H}\right]}}$$

where N_G and N_H are the numerators of the term a_{XY} for each of the populations G and H. The terms a_G and a_H are the estimates of the term a_{XY} for each population. If $T < 2$ then there is no difference between the two populations. If $T > 2$ then the two populations differ significantly.

Summary

The purpose of the present chapter is to illustrate some of the principles which may be used to quantify biological problems and answers and to outline the philosophy of the subject rather than to provide a scholarly or comprehensive teaching text. A large amount of interesting work has been ignored and any experienced worker glancing at the bibliography may be surprised at what may appear to be remarkable omissions.

That biology is becoming more quantitative cannot be denied. Many more biologists than previously are attempting to formulate and solve their problems using some degree of mathematical reasoning. Stereology is playing an increasingly important role as part of that quantitative approach. It is envisaged that there will be a considerable extension of the current interest in stereology into many other disciplines, for example pathology. It seems likely that many more stereological principles and techniques will be described. Some of these, together with some of those currently available, will doubtless become the basis for the construction and use of even more complex electronic devices for their practical application.

References

Cruz, L. M. (1976). Quantifying pattern. A stereological approach. *Journal of Microscopy*, **107**, 1–18.

DeHoff, R. T. & Rhines, F. N. (1968). *Quantitative Microscopy*. McGraw-Hill Book Company, New York.

Delèsse, A. (1848). Pour déterminer la composition des roches. *Annales des Mines, Série* 4, **13**, 379–88.

Elias, H., Hennig, A. & Schwartz, D. E. (1971). Stereology: applications to biomedical research. *Physiological Reviews*, **51**, 158–200.

Hally, A. D. (1964). A counting method for measuring the volumes of tissue components in microscopical sections. *Quarterly Journal of Microscopical Science*, **105**, 503–17.

Hennig, A. & Elias, H. (1971). A rapid method for the visual determination of size distribution of spheres from the size distribution of their sections. *Journal of Microscopy*, **93**, 101–8.

Mayhew, T. M. & Cruz, L. M. (1973). Stereological correction procedures for estimating true volume proportion from biased samples. *Journal of Microscopy*, **90**, 287–99.

Underwood, E. E. (1968). Stereology, or the quantitative evaluation of micro-structures. *Journal of Microscopy*, **89**, 161–80.

Underwood, E. E. (1970). *Quantitative Stereology*. Addison-Wesley Publishing Company, Reading, Massachusetts.

Weibel, E. R. (1972). A stereological method for estimating volume and surface of sarcoplasmic reticulum. *Journal of microscopy*, **95**, 229–42.

Weibel, E. R. & Bolender, R. P. (1973). Stereological techniques for electron microscopic morphometry. In *Principles and Techniques of Electron Microscopy, Biological Applications*, ed. M. A. Hayat, vol. 3, pp. 237–96. Van Nostrand Rheinhold Co. Ltd, London.

Weibel, E. R., Kistler, G. S. & Scherle, W. F. (1966). Practical stereological methods for morphometric cytology. *Journal of Cell Biology*, **30**, 23–38.

R. W. HORNE

Optical diffraction analysis of periodically repeating biological structures

It has long been recognised that the potential resolution of 0.2 to 0.3 nm from high-performance electron microscopes using non-biological specimens far exceeds the information obtainable from biological materials. Recent studies have indicated that the relatively poor average performance of 2.5 to 3.0 nm from biological specimens can be attributed to several difficult problems. First, preservation of specimen structure at the molecular level is limited by the existing preparative techniques available to the biologist. The number of fixatives and staining procedures for the preparation of materials for thin sections or as isolated particles is still relatively small, and only a few advances have been made in recent years to develop and improve methodology. Secondly, there is the problem of radiation damage to the specimen when it is illuminated by the electron beam, with consequent structural changes and the possible redistribution of heavy metal stains associated with the specimen. In addition, there is a rapid build-up of carbonaceous deposits at the specimen area illuminated by the electron beam.

In any electron optical system a certain amount of 'noise' is present in the final image. This is generated by the random arrival of electrons, amorphous background material distributed over the specimen, focusing effects and background levels in the photographic plate or film. The signal-to-noise ratio in the electron microscope image can also be influenced by the thickness of the specimen coupled with the physical dimensions of the biological structure in relation to the density and thickness of its support film.

When all of the above problems are added together it is not surprising that the average visible detail from biological specimens observed directly in the electron microscope cuts off at about 2.5 to 3.5 nm. For a more detailed discussion relating to these problems the reader is referred to the publications of Cosslett (1956), Dowell (1963), Kobayashi & O'Hara (1966), Yada & Hibi (1968), Stenn & Bahr (1970), Williams & Fisher (1970), Riecke (1971), Glaeser (1971), Horne (1973), Hart & Yoshiyama (1975) and others.

Several new experimental procedures have been devised in an attempt to overcome some of the basic problems mentioned above and these will be discussed below. As these methods are quite complex only a few basic examples together with results have been included in this article as an indication of the current progress.

Enhancement of periodic image detail

In recent years several new experimental techniques have been developed in an attempt to overcome problems associated with the signal-to-noise ratio in the final electron microscope image. Much of the progress has come from the important observation that many biological systems are assembled from relatively small structural components to form regularly arranged or symmetrical designs. The presence of periodic or symmetrical patterns associated with many biological particles recorded on electron micrographs can be used to advantage as it allows some reinforcement of the features appearing as repeating image components. Markham *et al.* (Markham, Frey & Hills, 1963; Markham, Hitchborn, Hills & Frey, 1964) developed a direct photographic method for reinforcing both linear and rotational periodic structures observed in viruses. The technique consisted of photographically superimposing objects at the correct linear distances, or in the case of rotational symmetry at the appropriate angles. In the original studies by Markham *et al.* (1963, 1964), the values required for the correct superimposition distances were determined by trial and error, but when used in parallel with optical diffraction techniques as discussed below, a more accurate calculation for the repeating distances was made possible.

An important development in the analysis of periodicities recorded in electron micrographs was the application of the optical diffractometer originally used by Taylor & Lipson (1951, 1964) in their studies of X-ray diffraction diagrams and related problems of interpretation. The basic system is shown in Fig. 1 and illustrates how the diffracted rays from an electron micrograph (or other suitable periodic object) can be brought to focus and produce a diffraction pattern. The distances of the spots from the centre of the diffraction pattern will be the inverse of the spacings or periodicities present in the original image. Assuming that some care has been taken in the calibration of the magnification at which the plate or film was recorded and the camera length of the diffractometer, images can be quantitatively analysed by measuring the distances, angles and distribution of the spots. Obviously the light spectra appearing in such a diffraction pattern or Fourier transform can be used to determine the spacings and angles for the purposes of photographic averaging of the same image as mentioned above (Markham *et al.*, 1963, 1964).

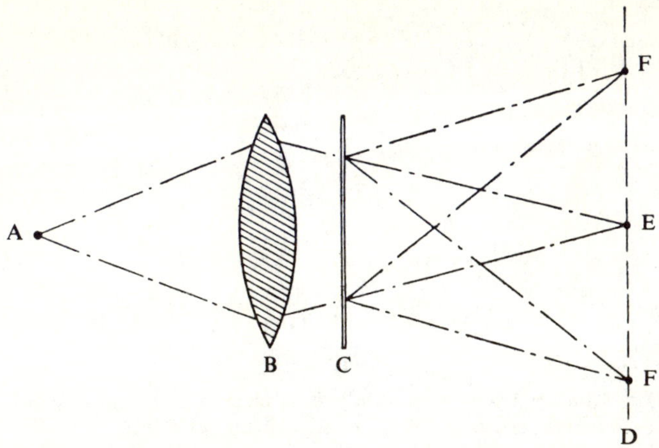

Fig. 1. The diagram illustrates the formation of an optical diffraction pattern from an electron micrograph containing periodic information. Monochromatic light is provided from the source at A. The diffraction lens B, forms a diffraction pattern in the plane D, from the electron micrograph C. Periodic detail recorded in C will produce discrete spots or spectra, E and F. Amorphous detail in C will also appear as a random disc of 'noise' superimposed on the spectra E and F (see Fig. 3).

The application of the optical diffractometer to electron micrographs was described by Klug & Berger (1964) and Finch, Klug & Stretton (1964) and has since developed into an important analytical method for electron microscopists. The basic arrangement of an experimental diffractometer is illustrated in Fig. 2 and can be constructed from standard optical laboratory apparatus (Markham, 1968; Horne & Markham, 1973). It will be seen from the diagram that in addition to the formation of a diffraction pattern, there is the possibility of reconstructing the image optically from the diffraction spots. The system can be considered as a type of spatial filter where the signal appears as a regular distribution of diffracted spots over a disc of random noise, the latter being generated from the non-periodic components in the original image (Fig. 3). Image reconstruction is achieved by making a suitable mask or filter consisting of holes or slits of the correct size to allow the diffracted spots to be imaged by a reconstruction lens (Fig. 2).

Investigations on the optical reconstruction of images from electron micrographs have been mainly associated with periodic detail observed in negatively stained specimens of viruses or their components (see Bancroft, Hills & Markham, 1967, Finch & Holmes, 1967; Erickson & Klug, 1971; DeRosier, 1971; Klug, 1971; Moody, 1971, Horne & Markham, 1973). It follows that the detail recorded in the electron micrographs and subsequent optical

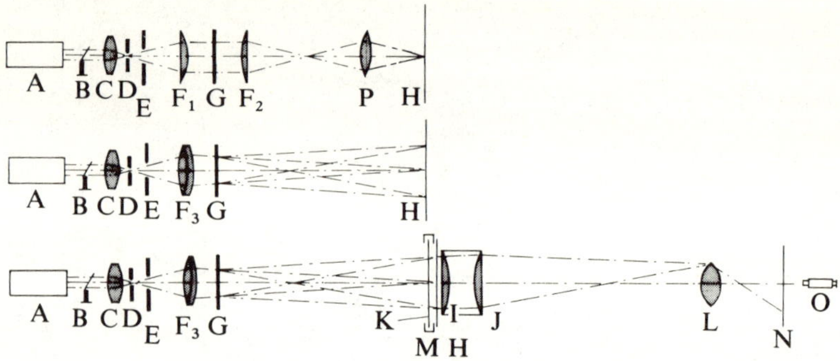

Fig. 2. The diagram illustrates the practical optical arrangements for a diffractometer used in the analysis of electron micrographs and image reconstruction. A, 2-milliwatt helium neon laser, λ = 632.8 nm. B, shutter for photographic exposure of the optical diffraction pattern or reconstructed image. C, beam expanding lens. D, pinhole to cut out any extraneous light from the laser (sometimes referred to as a side-band filter). E, adjustable diaphragm for limiting the illuminated area of the micrograph to about 1–2 cm diameter. F_1 and F_2, Collimating lenses to form a parallel beam of light incident on the micrograph. The F_2 lens may be omitted and then the micrograph is placed directly between the F_3 lens and the screen, H, as in the simple diffractometer illustrated in Fig. 1.
G, the electron micrograph, or optical grating or test object.
H, the diffraction plane, in which may be placed: (*a*) a ground glass screen for viewing the diffraction pattern; or (*b*) a camera system (preferably a Polaroid Land camera back) for recording the diffraction pattern; or (*c*) a filtering mask, K, in its holder, M. (The planes of H and K are shown separated in the diagram for illustration purposes only.) I and J, image reconstruction lens system used to form an image in the plane, N, with the beams allowed to pass through the filtering mask, K. K, filtering mask. L, magnification lens to enlarge the reconstructed image to a convenient size for recording. This lens may be omitted if sufficient optical bench space is available to allow a long throw onto the final viewing screen or camera at N.
M, filtering mask holder, made to allow accurate alignment of the mask, K, with the recorded position of the diffraction pattern.
N, final reconstructed image plane, in which may be placed: (*a*) a ground glass screen for observing the reconstructed image; or (*b*) a camera system for recording the reconstructed image. O, viewing telescope to allow critical adjustment of the optical system so that the reconstructed image is focused on the final viewing screen, at N. P, magnification lens to enlarge the optical diffraction pattern, if required. (From Horne & Markham, 1973.)

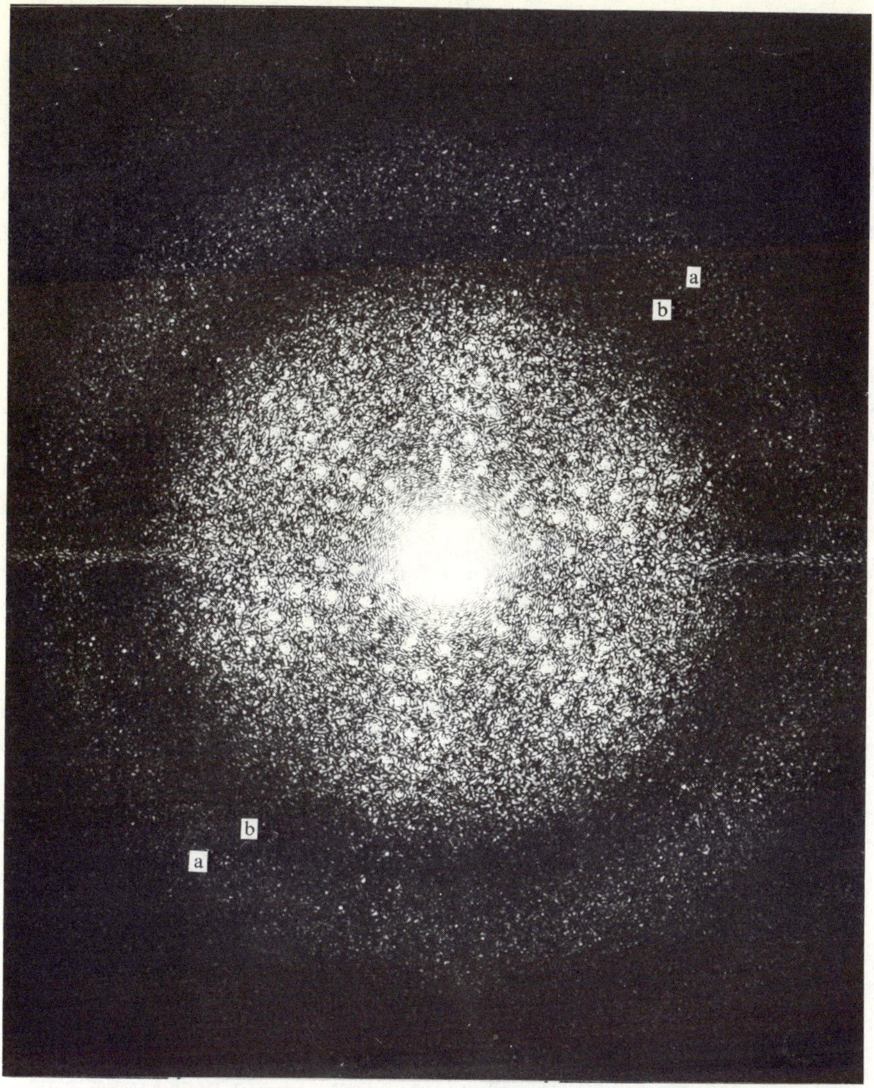

Fig. 3. Optical diffraction pattern obtained from an electron micrograph of turnip rosette virus in crystalline array. The spectra from the regular lattice structure will appear as the discrete spots in the pattern and the random noise as a circular disc of diffuse spectra. The diameter of the noise disc, a–a, is determined by the degree of underfocusing in the original electron microscope image. A zone of phase reversal is indicated at b–b in the underfocused micrograph producing the diffraction pattern. Defects in the electron image such as astigmatism will produce an elliptical area of noise in the diffraction pattern seen as loss of spectra in one direction. Although the pattern shown is underfocused, the astigmatism has been corrected to within reasonable limits and shows a symmetrical disc on noise and diffraction spectra. (Horne, Hull & Hobart, unpublished data.)

diffraction patterns is dependent on the penetration of negative stain coupled with preservation of the specimen and phase contrast effects introduced by the degree of defocusing in the electron microscope objective lens. Moreover, any defects in the photographic plate or film will also appear in the diffraction pattern (Erickson & Klug, 1971). For these reasons the reconstructed image may not necessarily represent the structural detail present in the original specimen and some additional independent experimental data are often required to support the features present in the final image.

A further and important development has been the quantitative three-dimensional reconstruction of structures from electron micrographs (De Rosier & Klug, 1968; Crowther, De Rosier & Klug, 1970; Crowther & Klug, 1975). The method necessitates recording different views of the specimen (i.e. tilting at known angles or viewing similar particles at different orientations within the same field), and the optical diffraction patterns from the micrographs provide a guide to the optimum detail and focusing requirements, etc. Each electron micrograph corresponding to a different view of the specimen is scanned optically to provide a digitised image. These optical data appear as a set of numbers for computer processing and are suitably programmed to provide a calculated Fourier transform. In the case of small viruses three or four views are sufficient to allow a three-dimensional Fourier inversion to be plotted section by section. The final three-dimensional image is built up from the plotted contours (Crowther & Amos, 1971*a*). In addition to the three-dimensional reconstruction, the computer processing of electron micrographs with rotational symmetry has been successfully performed (Crowther & Klug, 1975).

This development of computer programmes for electron microscope image analysis provides a more objective and quantitative assessment of the structural features present and shows whether sufficient data have been extracted from the image to produce reliable averaging or image reconstruction. However, it should be mentioned at this point that many of the structures analysed by all of the above methods are relatively simple, and often from single biological particles. With the development of new techniques for the isolation and preparation of specimens with highly complex periodic architecture, the analysis and image reinforcement of electron micrographs from specimens of this type (Roberts & Hills, 1976) may require even more sophisticated methods for extracting data and computer programmes (Fig. 4).

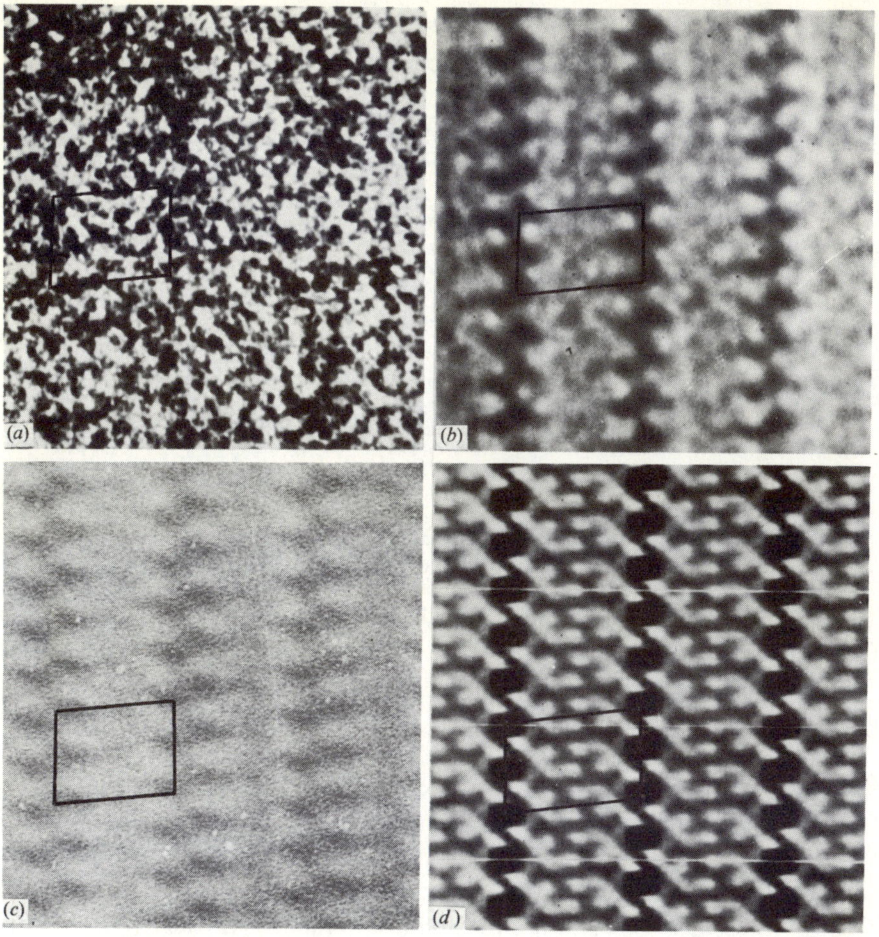

Fig. 4. This series of pictures illustrates the application of image analysis and reconstruction techniques to the crystalline cell wall of *Chlorogonium*. (*a*) Part of an original electron microscope image of a negatively stained *Chlorogonium* cell wall. × 900000. (*b*) Reconstructed image from the optical diffraction pattern obtained from (*a*), using the diffractometer illustrated in Fig. 2. × 1000000. (*c*) Image from a direct photographic linear integration of (*a*). × 900000. (*d*) Computer-reconstructed image of (*a*). The image was obtained by computing the autocorrelation function of the amplitudes in the electron diffraction and optical diffraction pattern of the in-focus image. The four areas marked in (*a*) to (*d*) indicate a primitive unit cell in the lattice structure. (Reproduced by kind permission of K. Roberts & G. J. Hills.)

Low-angle electron diffraction

The use of selected area low-angle electron diffraction for the study of inorganic crystalline materials has been well established. In the case of biological crystalline or paracrystalline structures where the periodicity is often in the range of 3 to 100 nm special problems are encountered with particular reference to the electron microscope optical system and camera lengths used in the low-angle electron diffraction mode (Mahl & Weitsch, 1961; Bassett & Keller, 1974; Ferrier & Murray, 1966; Ono, Ibe & Yotsumoto, 1960; Agar & Chescoe, 1974). One of the main limiting factors has been associated with extending the diffraction camera lengths of the electron microscope to values beyond those normally available in standard instruments. For many inorganic specimens where the interplanar spacings are of the order of 0.2 to 1.0 nm a camera length of about 40 to 100 cm is satisfactory. The camera lengths needed for biological materials, on the other hand, require values of about 10 to 200 m in order to separate the diffraction spectra at reasonable distances from the undiffracted beam recorded on the photographic plate (Ferrier & Murray, 1966; Murray, 1973).

The above technique is mentioned here because it has important future possibilities in being able to reconstruct images from the electron microscope diffraction patterns using similar computer methods to those currently being applied to image analysis and reinforcement of electron micrographs (Gerchberg, 1972; Gerchberg & Saxton, 1972). In addition, it is useful to compare the low-angle electron diffraction pattern directly with an optical diffraction pattern from the identical specimen area. It was mentioned earlier that the spectra appearing in the optical diffraction pattern are derived entirely from a photographic film or plate, and any physical defects, electron optical focusing effects and resulting contrast changes will also appear in the Fourier transform, but the low-angle electron diffraction pattern is formed directly from the specimen.

Several studies have also been made on the use of low-angle electron diffraction patterns to quantitate the effect of the electron beam on biological specimens. The comparison made between the resolution observed in electron microscope bright-field images and the corresponding low-angle electron diffraction patterns has raised a number of important questions about radiation damage and preservation of biological specimens under various electron dose rates (Glaeser, 1971). The examination of crystalline biological materials under conditions of low-angle electron diffraction enables the fading of the diffraction spectra to be measured at varying radiation dose rates and it is clear that there is some difference in sensitivity between various biological materials and stains during exposure to the electron beam (Glaeser, 1971). More recently

Unwin (1974) has shown that optical and computer analyses of bright-field electron micrographs have provided evidence for the redistribution and migration of uranyl formate negative stain over the surfaces of tobacco mosaic virus stacked disc protein.

It is clear from the low-angle electron diffraction studies of biological crystalline specimens that comparatively high orders of resolution of about 0.2 to 0.5 nm are obtainable where much lower levels of illumination can be used (cf. Glaeser, 1971; Matricardi, Moretz & Parsons, 1972; Unwin & Henderson, 1975). The problem of reconstructing images from complex electron diffraction patterns is a possibility, but a number of theoretical and technical problems have first to be resolved.

Specimen preparation

Although there is considerable current activity dealing with the electron microscope image, instrumentation and ancillary apparatus, relatively few investigations are in progress on the resolution limitations imposed by the specimen itself. The examination of bulk tissues or whole cells is still restricted to the existing methods of thin sections or freeze-fracture. Few of the materials prepared by these techniques show a resolution of better than about 2.5 to 3.0 nm, and for most of the investigations this level of structural information is considered to be adequate. The work on isolated biological particles, on the other hand, requires a much higher resolution if the morphological features of viruses, proteins and nucleic acids are to be more closely linked to other physical and biochemical data. Much of the electron microscopy associated with small biological structures at the quaternary structural level has come from negatively stained specimens and in many cases individual or small groups of particles were photographed.

The problems of resolving and measuring structure at the secondary and tertiary levels are closely linked to the limitations discussed in previous sections, but some improvement may come about from the specimen preparation procedures.

What is of importance here is the relation between the physical dimensions of the biological particles and the thickness of the film on which they are supported, coupled with the distribution of electron-dense stain at the film surface (Fig. 5). It is likely that very small irregularities at the support film surface will be filled with stain and the combination of a relatively thick film with randomly distributed areas of stain will decrease the signal-to-noise ratio in the image. It is also assumed that once the specimen is illuminated there will be an additional contamination layer building up over both support film

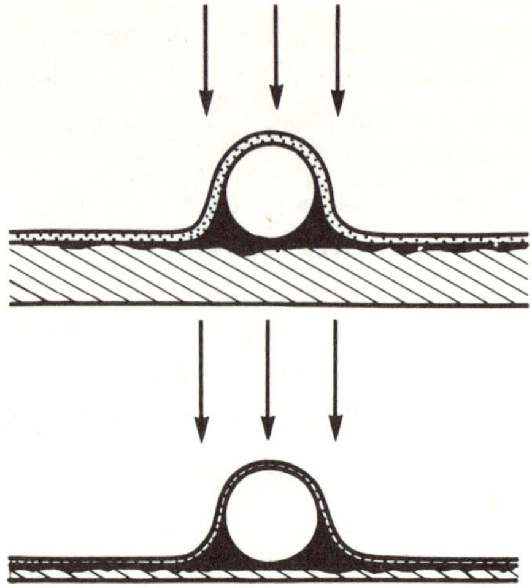

Fig. 5. (*a*) The diagram shows the size of a small biological particle of about 10 nm diameter in relation to the thickness of an average support film. In addition to the staining of the particle, small electron-dense areas will also appear at the film surface due to irregularities of the film or monolayers of low molecular weight material deposited at the time of preparation. During irradiation by the electron beam (arrows) a continuous layer of carbonaceous contamination is also building up over the specimen. Under the above conditions the signal-to-noise ratio in the final image is likely to be very low. (*b*) With a reduction in support film thickness, an even surface and low rate of contamination the signal-to noise-ratio can be improved, resulting in better resolution.

and paıticle, unless the carbonaceous contamination rate is reduced to a minimum. From our recent studies a contamination rate greater than 0.1 nm min^{-1} seriously affects the image detail at high resolution.

Experiments will be described now that provide evidence for an improvement in the signal-to-noise ratio brought about by developing specimen techniques suitable for image analysis and processing. They also indicate that it is possible to produce very thin carbon support films of more favourable dimensions for high-resolution work. Previous experimental negative staining procedures make use of depositing a droplet containing a mixture of virus, or other material, directly onto the specimen grid and support film (see Horne, 1965, 1967). In some experiments the negative stain and virus suspensions were allowed to extend across apertures in holey support films in order to reduce the total specimen thickness (Finch & Klug, 1966; Finch & Holmes, 1967). In the case of continuous support films the concentration of particulate material that can be deposited is obviously limited, unless very thick films are produced to maintain mechanical and thermal stability. The viruses completely embedded in stain extending across the holey film were

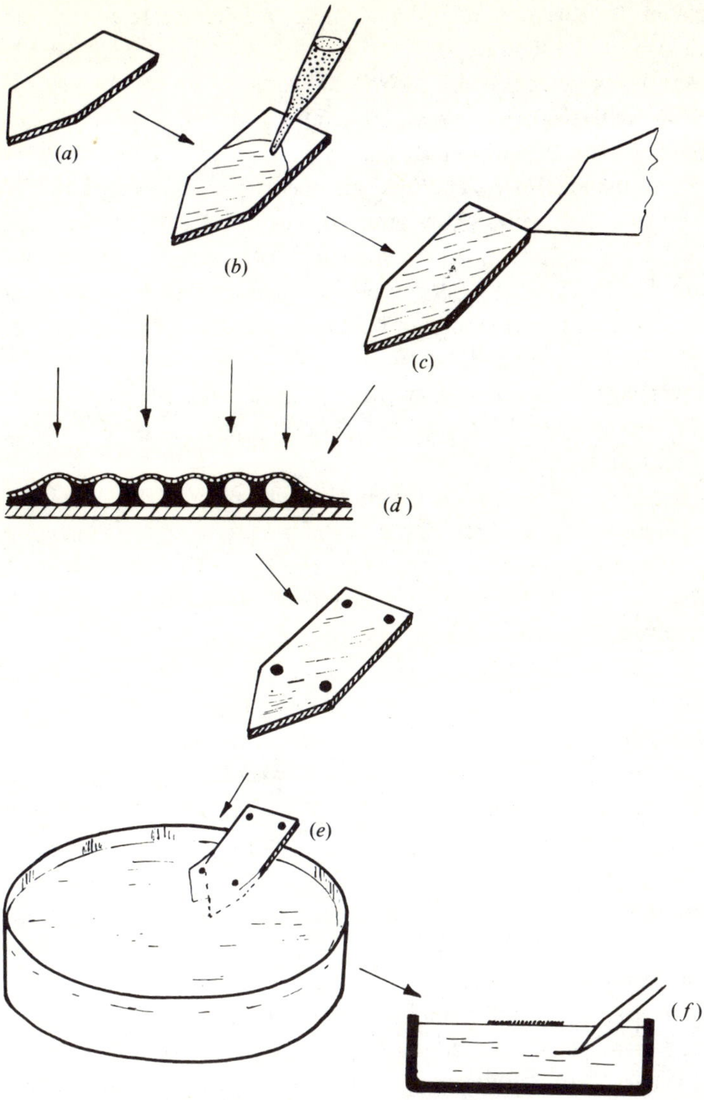

Fig. 6. The diagram shows the basic procedure for preparing specimens using the negative staining–carbon film method. A piece of mica (about 1 cm × 2 cm) is cut to the shape shown at (*a*), and split at the edge to provide two clean surfaces. The mixture of stain and virus material is spread over the surface (*b*), and drained with filter paper (*c*), leaving a wetted surface. The air-dried specimen is coated with a thin layer of carbon (*d*), and the specimen floated off at the air/liquid interface (*e*), of a second negative stain solution (0.5 % uranyl acetate). With the aid of a holey filmed grid the specimens are mounted as shown in (*f*).

also subject to thermal and electrostatic gradients, resulting in possible image drift and distortion.

The specimen preparation technique recently developed by Horne & Pasquali-Ronchetti (1974) for viruses and other small particles has special application to the resolution problems and quantitative image processing methods discussed above. The basic procedure is illustrated in Fig. 6. The results of the experiments show that: (*a*) very thin carbon films can be produced; (*b*) highly concentrated suspensions of virus (1 to 15 mg ml^{-1} virus) when mixed with negative stain can be mounted directly; and (*c*) suitable virus and other particle suspensions can be made to form crystals or paracrystalline arrays rapidly (Horne *et al.*, 1975*a*, *b*, 1976). Many of the crystalline arrays extend over large areas of 1 to 10 μm across and the packing of the particles within the arrays can be varied by changing the pH of the negative stain (Figs. 7 and 8).

Such specimens provided relatively large areas of highly ordered particles in two-dimensional or three-dimensional arrays and the images produced optical diffraction patterns of the type shown in Figs. 9 and 10. The virus crystalline arrays also provided some preliminary results from low-angle electron diffraction studies (Fig. 11).

Examination of the individual particles within the arrays showed that very few viruses presented images with the morphological protein units or capsomeres that are normally seen at their surfaces (Fig. 12); in particular those particles viewed along an axis of twofold rotational symmetry (cf. Bancroft *et al.*, 1967; Finch & Holmes, 1967). The optical diffraction spectra, on the other hand, contained spots extending out to a resolution of about 2.0 nm (Fig. 10). These marked differences in the appearance of the virus particles prepared by the negative staining-carbon method compared with those seen in conventionally prepared specimens could be explained on the basis that a more confused image results when the resolution is extended to a level where the individual protein structure units of molecular weight about 19000 are visualised (Fig. 12*a* and *b*). Assuming that the virus particles forming the square array shown in Fig. 8. were in the correct crystallographic lattice positions (i.e. rotated through 90° with respect to their neighbours), then it would be a simple matter to integrate the image diagonally superimposing the negative according to the method of Markham *et al.* (1963, 1964), and reinforce the symmetrical features of the virus particles. The result of this simple experiment is shown in Fig. 8(*b*) and the hexamer and pentamer morphological units are clearly visible (Fig. 12*a*). Obviously the entire lattice array is capable of reinforcement by multiple superimposition at the appropriate distances and angles.

Two important points emerged from these experiments. First, there was

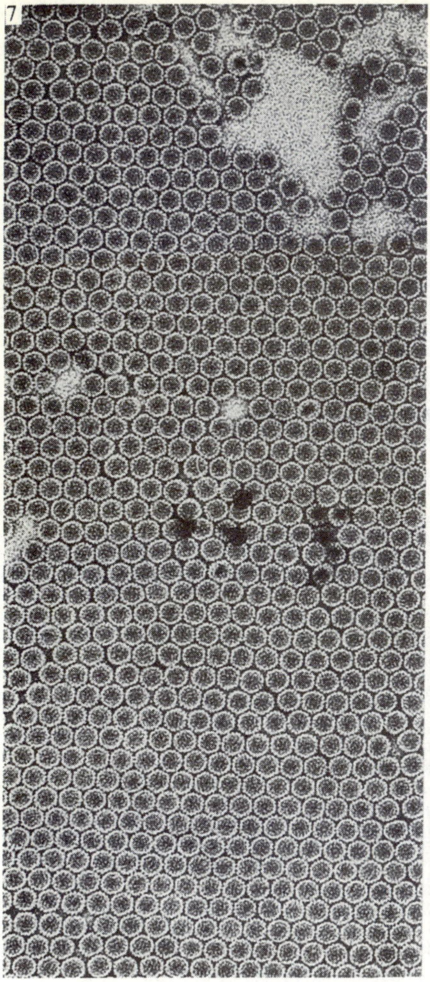

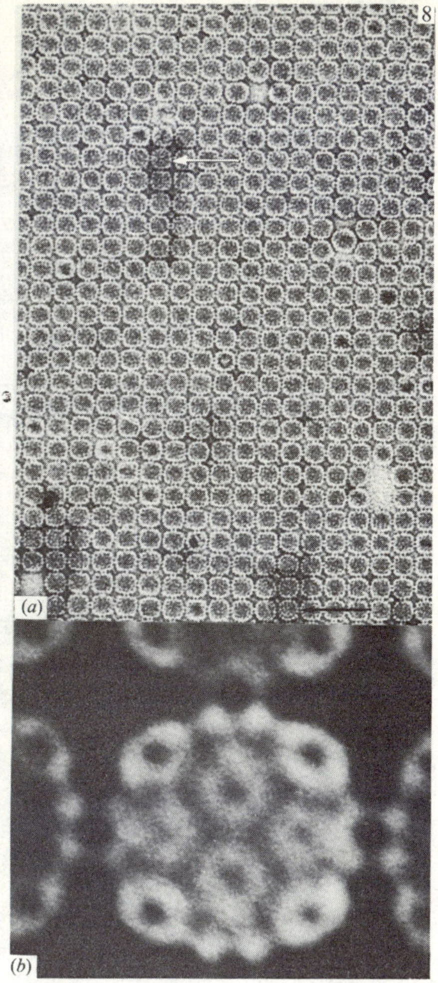

Fig. 7. Electron micrograph showing part of a two-dimensional array of cowpea chlorotic mottle virus in hexagonal array prepared by the method illustrated in Fig. 6. ×136000.

Fig. 8. (*a*) Cowpea chlorotic mottle virus particles shown in square arrays. ×136000. Few particles (arrowed) show the presence of capsomeres normally seen in conventionally negatively stained preparations. (*b*) The individual virus particles forming these arrays are rotated through 90° with respect to their neighbours, and by photographically integrating the imagc diagonally the capsomere distribution can be obtained. ×1 560000 (Horne *et al.* 1975*b*).

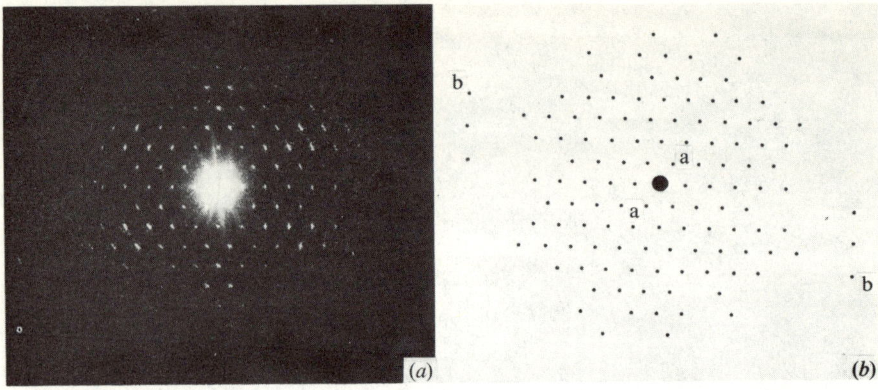

Fig. 9(*a*) and (*b*). The optical diffraction pattern and tracing showing the distribution of spectra in the original negative, was obtained from the array illustrated in Fig. 7. The first order spectra indicated at a–a correspond to the particle centre-to-centre distance of approximately 26 nm.

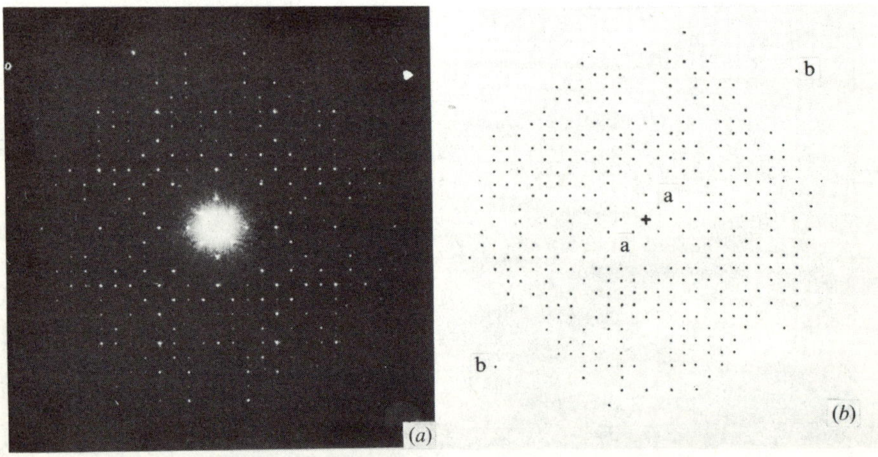

Fig. 10(*a*) and (*b*). Optical diffraction pattern and tracing from the square array of virus particles illustrated in Fig. 8 (a–a = 26 nm). The square arrays formed by cowpea chlorotic mottle virus were observed to give more complex optical diffraction patterns than the hexagonal arrays shown in Figs. 7 and 9 (Horne *et al.* 1975*b*).

structural information in the image that was sufficiently ordered to produce diffraction spectra of the type shown in Fig. 10; secondly, much of the disordered visible detail at higher resolution was confused with the background or noise level, but nevertheless cannot be excluded from possibly forming part of the virus particle or its components. This was indicated when the particles were examined at much higher magnifications where electron-dense crystallites of about 1.0 to 3.0 nm across were clearly seen associated

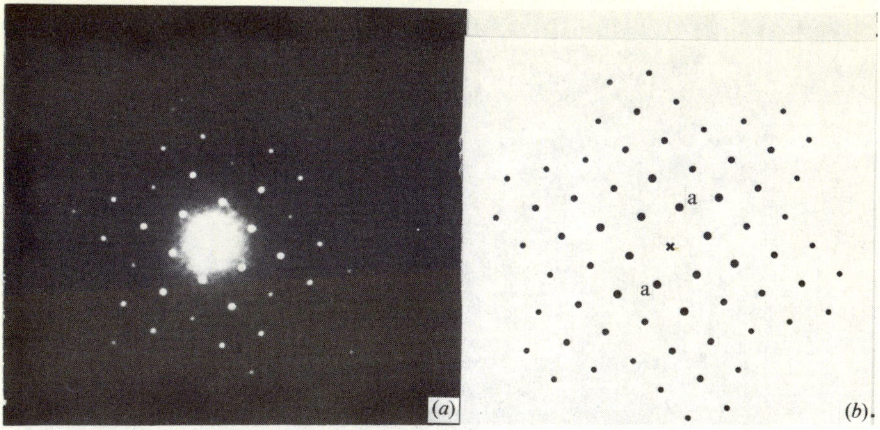

Fig. 11(*a*) and (*b*). Selected-area low-angle electron diffraction pattern and tracing obtained from a similar area to that in Fig. 7. The spectra were observed to be essentially similar to those seen in the optical diffraction pattern (see Fig. 9). Few commercial electron microscopes are fitted with electron optical systems able to provide camera lengths long enough to allow low-angle selected-area electron diffraction patterns to be obtained with relatively large lattice spacings of the dimensions shown in Figs. 9 and 10 (see text for details).

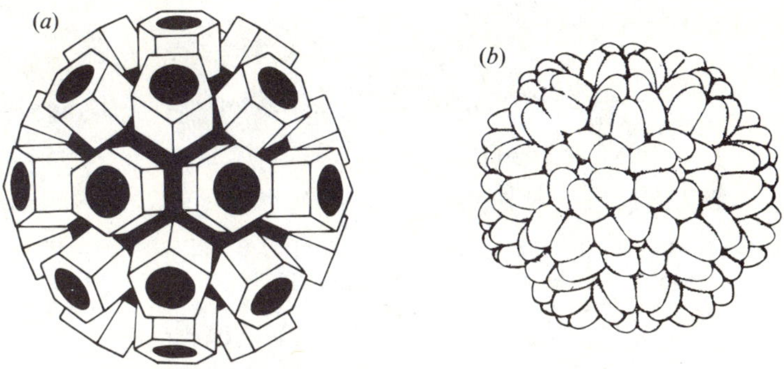

Fig. 12. (*a*) This drawing shows the distribution of twenty hexamer and twelve pentamer capsomeres or morphological units forming the T = 3 icosahedral protein shell of cowpea chlorotic mottle virus and structurally related particles. Electron microscope images from conventionally negatively stained specimens show structures possessing the same symmetry and arrangement as illustrated here. (*b*) The thirty-two capsomeres or morphological units are assembled from 180 protein structure units arranged as shown here. It can be seen that if the resolution is extended to visualise the structure units a more complex image results. (Fig. 12(*b*) is reproduced by kind permission of Dr L. D. L. Caspar, Brandeis University, Waltham, Mass., USA.)

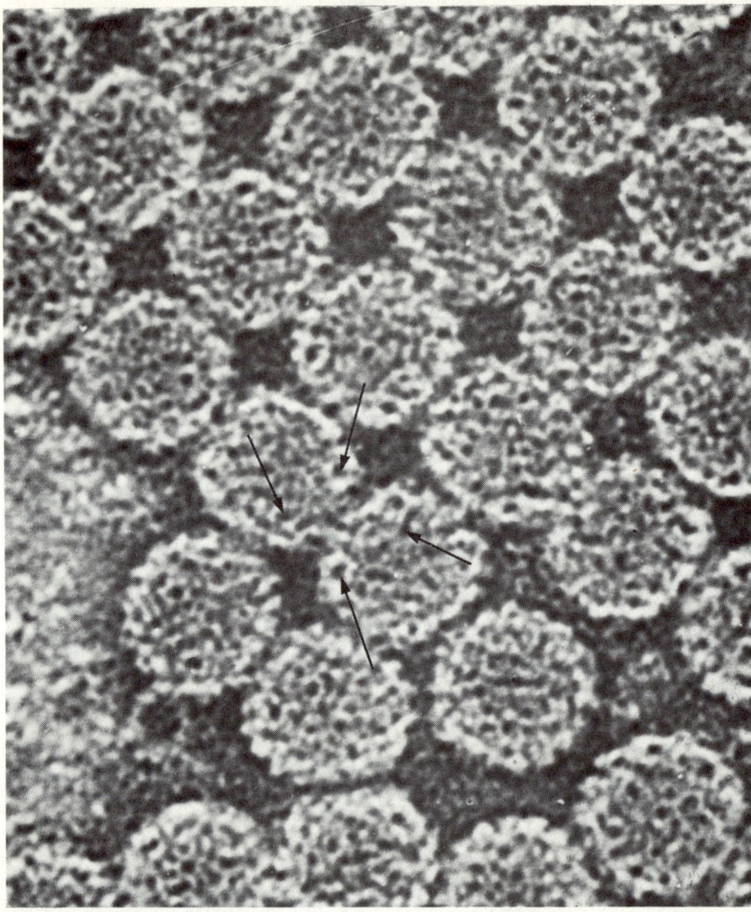

Fig. 13. Highly magnified region from a crystalline array of cowpea chlorotic mottle virus showing the distribution of small angular-shaped electron-dense crystallites (arrowed). The size of these small electron-dense particles was estimated to be within the range of 1.2 to 3.5 nm. ×776000.

with the virus protein surface (Fig. 13). These small crystals may be a stain–specimen composite formed as a result of radiation damage (Unwin, 1972, 1974).

In other experiments on filamentous and rod-like viruses using the negative staining-carbon method we obtained electron micrographs and optical diffraction spectra from aggregates of particles in roughly parallel arrays. The rods of tobacco mosaic virus serve as a good example to illustrate the application of optical diffraction and simple photographic averaging to produce an image that closely corresponds to the model derived from X-ray diffraction data (Figs. 14 and 15).

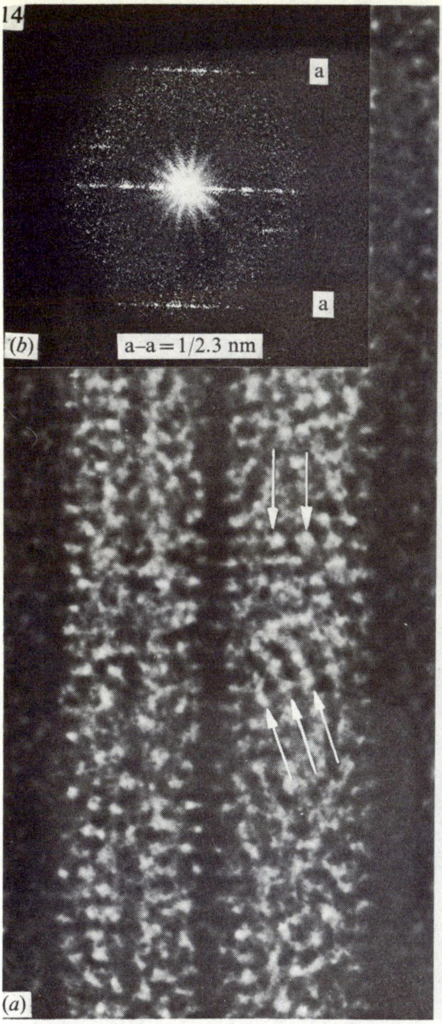

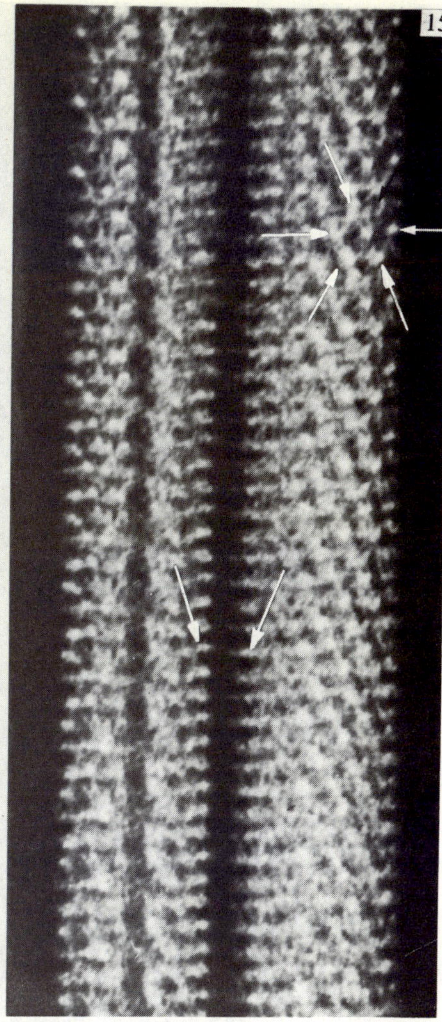

Fig. 14. (*a*) High magnification electron micrograph showing part of a tobacco mosaic virus rod. The 2.3-nm periodicity at the edges of the rod is visible against the background noise level, but few areas show the presence of protein structure units (arrows) forming the virus helix. ×1 045 000. (*b*) An optical diffraction pattern obtained from (*a*), showing the 2.3-nm (a–a) spacing corresponding to the periodicity of the protein primary helix and a repeat structure at about 6.9 nm along the rod axis.

Fig. 15. The distribution of structure units together with the primary and secondary helices can be reinforced by photographically integrating the image in Fig. 14(*a*) corresponding to a distance of 6.9 nm, or repeat distance of the primary helix at every three turns. ×1 045 000. (From Horne *et al.*, 1976.)

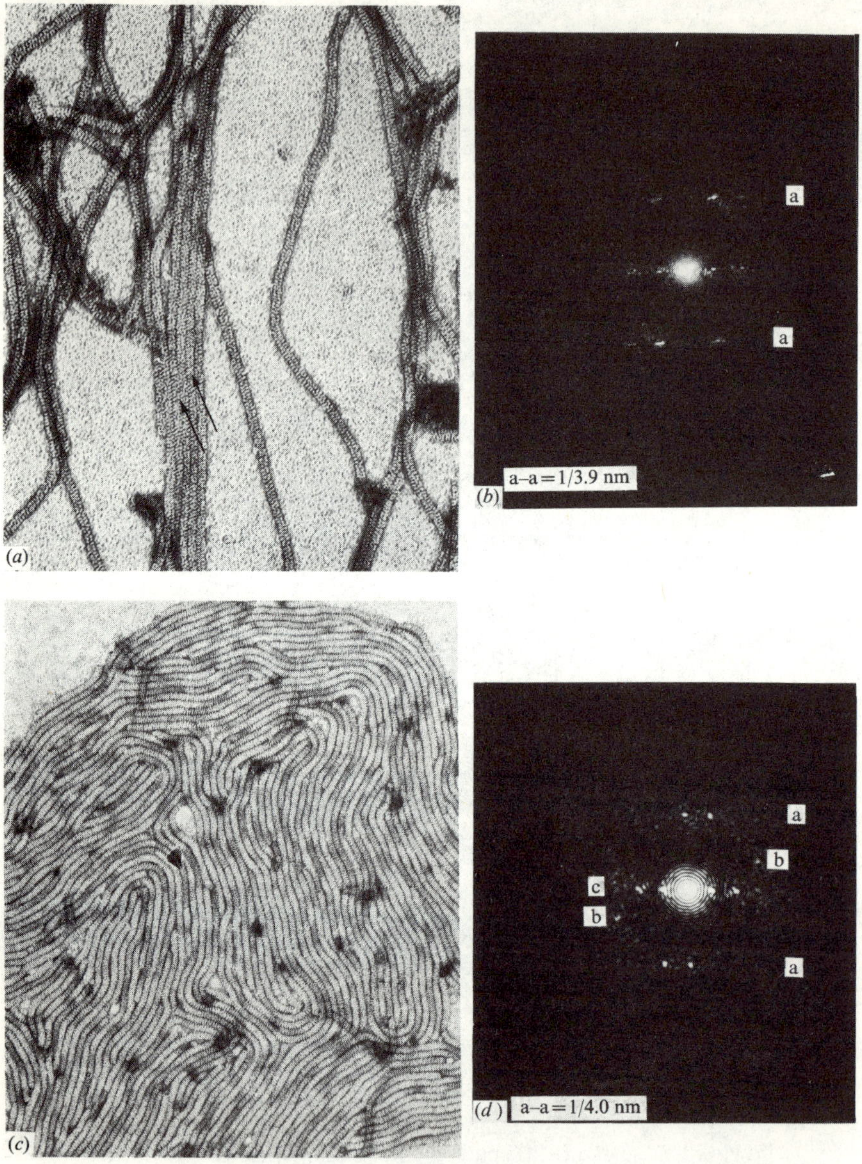

Fig. 16. (*a*) Flexuous filaments of sugar beet yellows virus negatively stained with ammonium molybdate showing well defined periodicity along the long particle axis. The angle of the striations across the filaments is reinforced where several particles are in parallel array (arrows). × 130000. (*b*) Optical diffraction pattern obtained from the parallel array of filaments shown in (*a*). The pitch of the primary helix forming the spectra a–a is 3.9 nm. (*c*) Aggregated filaments of apple chlorotic leafspot virus negatively stained with ammonium molybdate

The optical diffraction spectra from the flexuous viruses shown in Fig. 16 clearly indicate that the striations along each virus filament are derived from a helix and that there may be small differences in the pitch and diameter between certain groups of these viruses. The determination of such small differences by direct examination of the bright-field images would have been less reliable (Goodman, Horne & Hobart, 1975; Goodman, Macdonald, Horne & Bancroft, 1976; Horne *et al.*, 1975*b*).

An equally important application of optical diffraction and image analysis is the measurement of reconstituted virus products. When the proteins and nucleic acids of certain plant viruses are separated and purified from normal virus, they can be reassembled *in vitro* (Bancroft, 1970; Goodman *et al.*, 1975, 1976). These reassembed products can then be photographed in the electron microscope and their corresponding optical diffraction patterns recorded. The presence of any small structural changes in the reassembled products will appear in the diffraction pattern (Fig. 17). The reassembled protein alone from a filamentous virus is illustrated in Fig. 18(*a*) and the corresponding diffraction pattern shows the presence of a stacked-disc array instead of the normal helical array (Fig. 17*b*). Thus the optical diffractometer provides a critical test for comparative experiments of this type.

One interesting point to emerge from these studies was the presence of low molecular weight products in the virus suspensions and their effects on both crystal formation and image resolution. An appreciable amount of monomeric protein subunit material was observed in each of the freshly prepared virus suspensions we have examined. At high magnification this material was seen to form a monolayer in the specimen background and often extended over, or even embedded, the virus particles. The effect of this material was to present a confused image of the virus surface detail, and also to impair the formation of the large crystalline arrays.

In addition to the masking effects caused by both organic and inorganic materials forming monolayers at the specimen surface, there is the problem of three-dimensional preservation. Many biological structures in the form of isolated particles tend to distort or collapse during drying in the presence of negative stains. For the above reasons the dimensions and angles from models used to form simulated images of biological objects may show marked differences (Amos, 1974). The image shown in Fig. 15 was obtained from part

---

showing the flexuous structure of the virus and periodicity resulting from the helical structure. × 50000. (*d*) Optical diffraction pattern from (*c*). The spectra at c indicate the centre-to-centre distance of the closely packed filaments at 12.2 nm and the spectra a–a corresponding to the primary helical pitch of 4.0 nm. (From Horne *et al.* (1975). *Micron*, **5**, 233.)

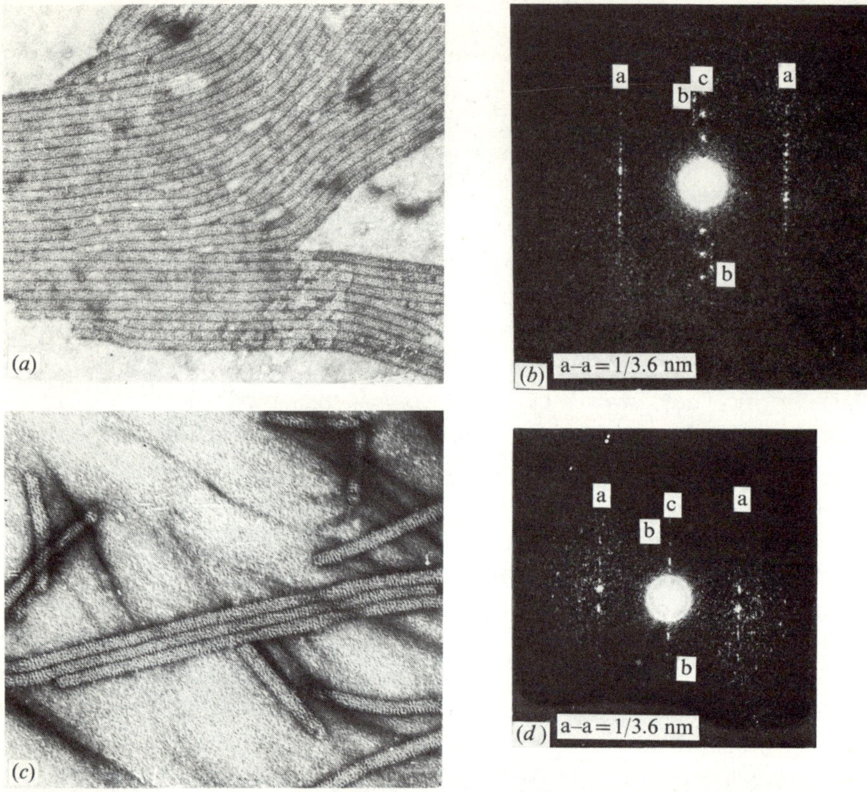

Fig. 17. (*a*) Filaments of native potato virus X packed with particles in approximate parallel arrays. Regular striations along the filaments are from the basic helical structure. × 125 000. (*b*) Optical diffraction pattern from (*a*) showing the spectra a–a derived from the primary helical pitch of 3.6 nm and the layer line at b–b indicating a repeat of the primary helix at every eight turns along the particle axis. The interparticle spacing at c is 12.6 nm. (*c*) Filaments reconstituted *in vitro* from purified potato virus X protein and RNA. The basic morphology of the reconstituted products is essentially the same as (*a*). × 132 000. (*d*) This optical diffraction pattern was obtained from the filaments in (*c*) and the spectra indicate that the reconstituted virus has the same structural organisation and dimensions as the native potato virus X shown in (*a*) and (*b*). (From Horne *et al.* (1975). *Micron* **5**, 233.)

of a tobacco mosaic virus rod photographed in the electron microscope and photographically integrated to reinforce the primary and tertiary helical lattices (Horne *et al.* 1976). It can be seen that the distribution of the protein structure units forming the cylindrical lattice is more clearly defined on one side of the particle than the other. This can be attributed to the possibility of the virus rod being tilted with respect to the incident electron beam, twisted about its long axis or partially flattened (Finch, 1972).

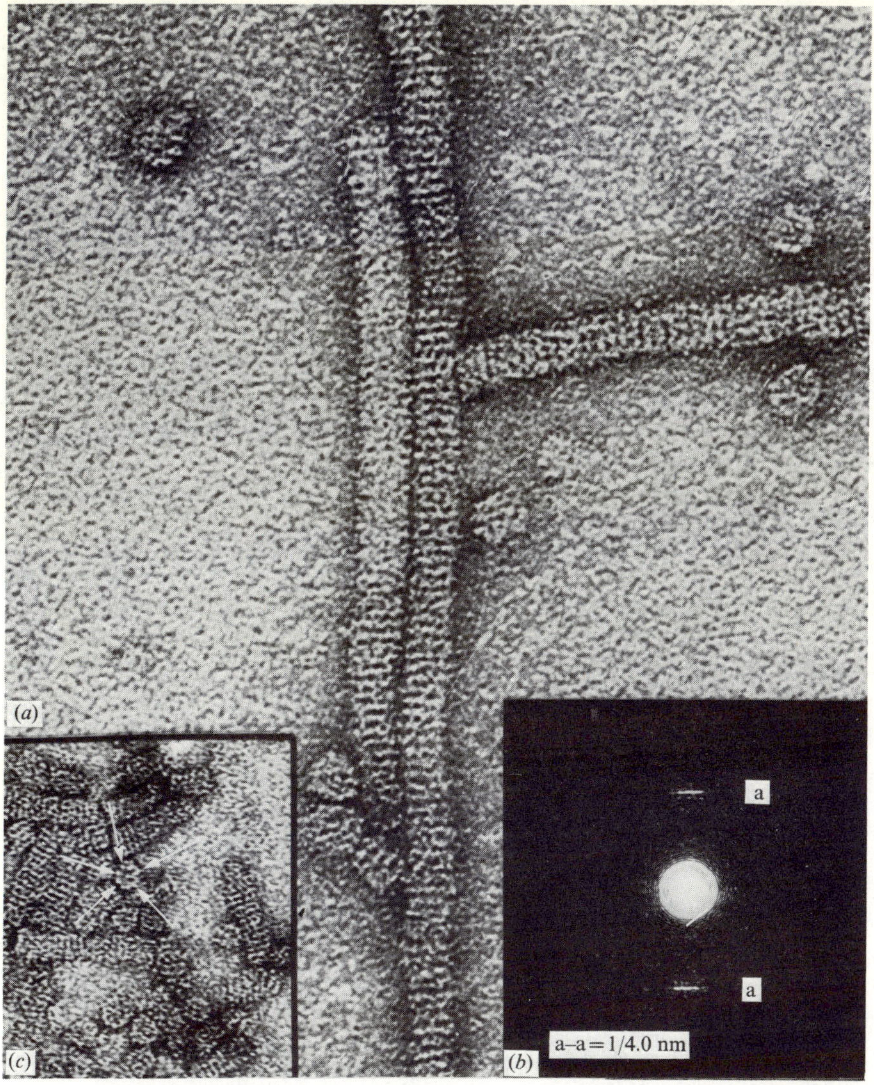

Fig. 18. (*a*) The filaments shown in the micrograph were from reconstituted protein purified from potato virus X. In the absence of nucleic acid (RNA) the structure formed *in vitro* was that of a stacked-disc arrangement. ×470000. (*b*) The position of the spectra a–a in the optical diffraction pattern shows the periodic spacing along the particle axis to be 4.0 nm and as a parallel arrangement of subunits which is consistent with a stacked-disc assembly. (*c*) Fragments of reconstituted potato virus X protein showing structures resembling subunits and particles viewed end on (arrows). ×265000. (From Goodman *et al.*, 1976.)

The recent use of glucose by Unwin & Henderson (1975) to preserve unstained biological crystals for analysis by low-angle electron diffraction has shown that a resolution of about 0.5 to 0.7 nm can be obtained. It is possible that the use of sugars and other compounds used in conjunction with new staining procedures could result in the production of very high resolution bright-field images. The problem of radiation damage may also be reduced by the use of preservative materials more resistant to electron beam effects and obviously requires further investigation.

Conclusions

The development of structural analysis of biological structures by image processing from electron micrographs provides a more quantitative approach to the interpretation of periodic information in the final image. With new techniques for isolating and purifying biological particles and macromolecules for electron microscopy, the application of these analytical and image reconstruction methods will play an important role in revealing the structural organisation of a wide range of specimens. In order to extend the limit of resolution more emphasis will have to be placed on the specimen preparation and preservation rather than the image processing techniques. (Crowther & Klug, 1975). Equally important is the need to develop similar image processing methods for non-periodic information in the electron microscope images of biological materials.

References

Agar, A. W. & Chescoe, D. (1974). *Principles and Practice of Electron Microscope Operation. Practical Methods in Electron Microscopy*, vol. 2, ed. A. M. Glauert. North Holland Publishing Co., Amsterdam.

Amos, L. A. (1974). Image analysis of macromolecular structures. *Journal of Microscopy*, **100**, 143–52.

Bancroft, J. B. (1970). The self assembly of spherical plant viruses. *Advances in Virus Research*, **16**, 99–134.

Bancroft, J. B., Hills, G. J. & Markham, R. (1967). A study of the self assembly process in a small spherical virus. *Virology*, **31**, 356–79.

Bassett, G. A. & Keller, A. (1964). Low angle scattering in an electron microscope application to polymers. *Philosophical Magazine*, **9**, 817–28.

Cosslett, V. E. (1956). Electron microscopy. In *Physical Techniques in Biological Research*, vol. 1, ed. A. W. Oster & G. Pollister, pp. 461–79. Academic Press, New York & London.

Crowther R. A. (1971). Procedures for 3-dimensional reconstruction of spherical viruses by Fourier synthesis from electron micrographs.

Philosophical Transactions of the Royal Society of London, Series B, **261**, 221–30.

Crowther, R. A. & Amos, L. A. (1971*a*). Three dimensional image reconstructions of some small spherical viruses. *Cold Spring Harbor Symposia on quantitative Biology*, **36**, 489–94.

Crowther, R. A. & Amos, L. A. (1971*b*). Harmonic analysis of electron microscope images with rotational symmetry. *Journal of Molecular Biology*, **60**, 123–30.

Crowther, R. A. & Klug, A. (1975). Structural analysis of macromolecular assemblies by image reconstruction from electron micrographs. *Annual Review of Biochemistry*, **44**, 161–82.

Crowther, R. A., De Rosier, D. J. & Klug, A. (1970). The reconstruction of a 3-dimensional structure from projections and its application to electron microscopy. *Proceedings of the Royal Society of London*, Series A, **317**, 319–40.

De Rosier, D. J. (1971). Three-dimensional image reconstruction of helical structures. *Philosophical Transactions of the Royal Society of London*, Series B, **261**, 209–10.

De Rosier, D. J. & Klug, A. (1968). Reconstruction of 3-dimensional structures from electron micrographs. *Nature*, *London* **217**, 130–4.

Dowell, W. C. T. (1963). Das Elektronenmikroskopische Bild von netznebenen Scharen und sein Kontrast. *Optik*, **20**, 553–68.

Erickson, H. P. & Klug, A. (1971). Measurements and composition of defocussing and aberrations by Fourier processing of electron micrographs. *Philosophical Transactions of the Royal Society of London*, Series B, **261**, 105–18.

Ferrier R. P. & Murray, R. T. (1966). Low angle electron diffraction. *Journal of the Royal microscopical Society*, **85**, 323–35.

Finch, J. T. (1972). The hand of the helix of tobacco mosaic virus. *Journal of Molecular Biology*, **66**, 291–4.

Finch, J. T. & Holmes, K. C. (1967). Structural studies of viruses. In *Methods in Virology*, vol. 3, ed. K. Maramorosch & H. Koprowski, pp. 351–474. Academic Press, New York & London.

Finch, J. T. & Klug, A. (1966). Arrangements of protein subunits and the distribution of nucleic acid in turnip yellow mosaic virus. II. Electron microscopic studies. *Journal of Molecular Biology*, **15**, 344–64.

Finch, J. T., Klug, A. & Stretton, A. O. (1964). The structure of the 'polyheads' of T4 bacteriophage. *Journal of Molecular Biology*, **10**, 570–5.

Gerchberg, R. W. (1972). Holography without fringes in the electron microscope. *Nature*, *London*, **240**, 404–6.

Gerchberg, R. W. & Saxton, W. O. (1972). A practical algorithm for the determination of phase from image and diffraction plane pictures. *Optik*, **35**, 237–41.

Glaeser, R. M. (1971). Limitations to significant information in biological electron microscopy as a result of radiation damage. *Journal of Ultrastructure Research* **36**, 466–82.

Goodman, R. M., Horne, R. W. & Hobart, J. M. (1975). Reconstitution of Potato Virus X *in vitro*. II. Characterization of the reconstituted product. *Virology*, **68**, 299–308.

Goodman, R. M., Macdonald, J. G., Horne, R. W. & Bancroft, J. B. (1976). Assembly of flexuous plant viruses and their proteins. *Philosophical Transactions of the Royal Society of London*, Series B, in press.

Hart, R. G. & Yoshiyama, J. M. (1975). Electron microscopy with reduced beam damage to the specimen. A retractable image intensifier. *Journal of Ultrastructure Research*, **51**, 40–5.

Horne, R. W. (1965). Negative staining methods. In *Techniques for Electron Microscopy*, 2nd ed. D. Kay, pp. 328–55. Blackwell, Oxford.

Horne, R. W. (1967). Some general remarks on the application of the electron microscope to the study of biological structure at the macromolecular level. In *Methods in Virology*, vol. 3, ed. K. Maramorosch & H. Koprowski, pp. 522–574. Academic Press, New York & London.

Horne, R. W. (1973). Contrast and resolution from objects examined in the electron microscope with particular reference to negatively stained specimens. *Journal of Microscopy*, **98**, 286–98.

Horne, R. W., Hobart, J. M. & Markham, R. (1976). Electron microscopy of tobacco mosaic virus prepared with the aid of negative staining–carbon film techniques. *Journal of general Virology*, **31**, 265–9.

Horne, R. W., Hobart, J. M. & Pasquali-Ronchetti, I. (1965*b*). A negative staining–carbon film technique for studying viruses in the electron microscope. III. The formation of 2-dimensional and 3-dimensional crystalline arrays of Cowpea Chlorotic Mottle Virus. *Journal of Ultrastructure Research*, **53**, 319–30.

Horne, R. W. Markham, R. (1973). Applications of optical diffraction and image reconstruction techniques to electron micrographs. In *Practical Methods in Electron Mircoscopy*, vol. 1, ed. A. M. Glauert, pp. 327–444. North-Holland Publishing Co., Amsterdam.

Horne, R. W. & Pasquali-Ronchetti, I. (1974). A negative staining–carbon technique for studying viruses. I. Preparative procedure for studying iocsahedral and filamentous viruses. *Journal of Ultrastructure Research*. **47**, 361–83.

Horne, R. W., Pasquali-Ronchetti, I. & Hobart, J. M. (1975*a*). A negative staining–carbon film technique for studying viruses in the electron microscope. II. Application to Adenovirus type 5. *Journal of Ultrastructure Research*, **51**, 233–52.

Klug, A. (1971). Applications of image analysis in electron microscopy. *Philosophical Transactions of the Royal Society of London*, Series B, **261**, 173–9.

Klug, A. & Berger, J. E. (1964). An optical method for the analysis of periodicities in electron micrographs, and some observations on the mechanism of negative staining. *Journal of Molecular Biology*, **10**, 564–9.

Kobayashi, K. & O'Hara, M. (1966). Voltage dependence of radiation damage to polymer specimens. *Proceedings of the 6th International Congress on Electron Microscopy, Kyoto*, **1**, pp. 579–80.

Mahl, H. & Weitsch, W. (1961). Kleinwinkelbeugune mit Elektron-enstrahlen. *Zeitschrift für Naturforschung*, **15a**, 1051–5.

Markham, R. (1968). The optical diffractometer. In *Methods in Virology*, vol. 4, ed. K. Maramorosch & H. Koprowski, pp. 503–29. Academic Press, New York & London.

Markham, R., Frey, S. & Hills, G. J. (1963). Methods for the enhancement of image detail and accentuation of structure in electron microscopy. *Virology*, **20**, 88–102.

Markham, R., Hitchborn, J. H., Hills, G. J. & Frey, S. (1964). The anatomy of tobacco mosaic virus. *Virology*, **22**, 342–59.

Matricardi, V. R., Moretz, R. C. & Parsons, D. F. (1972). Electron diffraction of wet proteins:catalase. *Science*, **177**, 268–70.

Moody, M. F. (1971). Applications of optical diffraction to helical structures in the bacteriophage tail. *Philosophical Transactions of the Royal Society of London*, Series B, **261**, 181–95.
Murray, R. T. (1973). The possibilities for electron diffraction in biology. *Journal of Microscopy* **98**, 345–51.
Ono, A., Ibe, K. & Yotsumoto, H. (1970). On a low-angle scattering electron diffraction. *Journal of Electron Microscopy* **19**, 149–55.
Riecke, W. D. (1971). Prospects foı high resolution electron microscopy. *Philosophical Transactions of the Royal Society of London*, Series B, **261**, 15–34.
Roberts, K. & Hills, G. J. (1976). The crystalline glycoprotein cell wall of the green alga *Chlorogonium elongatum*: a structural analysis. *Journal of Cell Science*, **21**, 59–71.
Stenn, K. & Bahr, G. F. (1970). Specimen damage caused by the beam of the transmission electron microscope, a correlative reconsideration. *Journal of Ultrastructure Research*, **31**, 526–50.
Taylor, C. A. & Lipson, H. (1951). Optical methods in crystal structure determination. *Nature, London*, **167**, 809–10.
Taylor, C. A. & Lipson, H. (1964). *Optical Transforms*. G. Bell & Sons Ltd, London.
Unwin, P. N. T. (1972). Negative staining of biological specimens using mixture salts. In *Proceedings of the 5th European Congress on Electron Microscopy*, pp. 232–3. Institute of Physics, London.
Unwin, P. N. T. (1974). Electron microscopy of the stacked disk aggregate of tobacco mosaic virus protein. II. The influence of electron irradiation on the stain distribution. *Journal of Molecular Biology*, **87**, 657–70.
Unwin, P. N. T. & Henderson, R. (1975). Molecular structure determination by electron microscopy of unstained crystalline specimens. *Journal of Molecular Biology*, **94**, 425–40.
Williams, R. C. & Fisher, H. W. (1970). Electron microscopy of tobacco mosaic virus under conditions of minimal beam exposure. *Journal of Molecular Biology*, **52**, 121–3.
Yada, K. & Hibi, T. (1968). Influence of illumination beam on the image contrast. In Proceedings of the 4*th European Congress on Electron Microscopy*, vol. 1, ed. D. Steve Bocciarelli, pp. 143–4. Tipografia Poliglotta Vaticana, Rome.

J. S. PLOEM

Quantitative fluorescence microscopy

In many fields of histochemistry and cytochemistry, quantitative fluorescence microscopy has become a valuable technique. The method is especially sensitive if only low concentrations of reaction products are present in a microscopical specimen. In general the instrumentation required is simpler than that needed for absorption methods, since no scanning in the object or image plane is required to avoid a distributional error occurring with the densitometric measurement of inhomogeneously distributed chromophores (Ornstein, 1952; Mayall & Mendelsohn, 1970).

Microfluorometry has been used mainly to measure fluorescent reaction products (FRPs) in fixed tissues and cells. Microfluorometric methods are also suited for kinetic studies if the FRP is formed in the living cell. Microfluorometry is an essential method for cytochemical analysis of cells in suspension in flow-through systems. Using a specific staining, for example of DNA, protein or components on the cell membrane, a large number of cells can be quantitatively analysed for several cellular parameters, or even separated into several subpopulations of cells if the recently developed fluorescence-activated cell sorters are used (Fulwyler, 1974).

In recent years microfluorometry has also been used in microchemistry, for the quantitation of FRPs in extremely small volumes of fluid contained in microdroplets, microcapillaries or microspheres of, for instance, Sepharose (Ploem, 1975). The increase in applications of microfluorometry in the past two decades has been followed by rapid technical developments in microfluorometers. Both simple and very sophisticated instruments have become available.

Recent developments in electronics and the considerable drop in price of small digital computers have contributed to the construction of the more versatile instruments. The availability on the market of such automated instruments has in turn led to an increased use of microfluorometry in those routine and research applications where microfluorometry with a conventional instrument was too time-consuming.

Fluorescence microscopy

Fluorescence microscopy is based on the property of some substances to absorb light in a certain wavelength range. This energy is subsequently partly emitted in the form of light. In most cases the wavelength of the emitted light is longer than that of the absorbed light. The fluorescence microscope therefore has to be constructed in such a way that the shorter wavelengths of the excitation light are separated from the longer wavelengths of the emitted fluorescence light. Since the intensity of the emitted fluorescence is often relatively weak when compared to the intensity of the strong exciting radiation, it is absolutely necessary to prevent the mixing of exciting radiation with the emitted fluorescence light in the microscopical image.

The fluorescence microscope must fulfil two main functions: (1) it must excite the FRP in the microscopic specimen effectively, preferably with wavelengths close to the absorption peak of the FRP; and (2) it must collect as much as possible of the fluorescence light so as to optimise its visibility in the microscope.

The quality of the microscopical image in fluorescence microscopy depends mainly on the image contrast and the image brightness. The image contrast is determined by the ratio of the fluorescence emission of those structures of the specimen which are specifically stained to the light observed in the background. The latter is composed of unwanted excitation light still transmitted by the secondary filters used to stop the excitation light, and the autofluorescence of tissue components and the optical parts of the microscope. All efforts in fluorescence microscopy are directed towards obtaining an optimal image contrast, while maintaining a sufficiently bright image. It is evident that measures to obtain an optimal image contrast may result in very dim fluorescence images, due for instance to the use of filters with extremely narrow transmission bands which have often only a relatively low transmittance. A sufficiently bright image is required for easy visual observation and microphotography with reasonably short exposure times. Therefore excitation with wider transmission bands is often required, which may lead to fluorescence of not only the specific FRP but also of other tissue components and thus to non-specific background fluorescence. A compromise between the intensity of specific fluorescence and a reasonably low level of non-specific fluorescence in the background must in most applications be accepted.

For review articles on fluorescence microscopy the reader is referred to articles by Young (1961), Barer (1964), Goldman (1968), Nairn (1969), Walter (1970), Rost (1972) and Kraft (1975).

Main components of a fluorescence microscope

The light necessary for the excitation of a fluorochrome is obtained from a light source with an arc or filament of high intrinsic brilliance. High-pressure mercury or xenon arcs and halogen–quartz lamps are mostly used for this purpose. For excitation with ultraviolet and violet light high-pressure arc lamps must be used since halogen–quartz lamps emit excitation light of sufficient intensity for fluorescence microscopy only in the blue and green wavelength range above 450 nm (Tomlinson, 1967, 1971).

All light not wanted for the excitation of the FRP must be removed from the exciting light beam using a primary filter. This filter, called the excitation filter, must absorb or reflect all unwanted light. Coloured glass filters have been commonly used for this purpose, but interference filters have recently been increasingly used. The concentration of the excitation light on the microscopical specimen is achieved by using a light collecting system such as a substage condenser for transmitted illumination and an annular condenser or an objective for epi-illumination.

After the excitation light has reached the microscopical specimen, it will often still enter the objective in the direction of the eyepieces, owing to scattering or reflection in the specimen. This unwanted excitation light must be absorbed or reflected by a secondary filter. This filter, called the barrier filter, prevents the unwanted excitation light from reaching the eyes of the observer, or the photomultiplier in the case of a microfluorometer. The secondary filter must have a high transmittance for the fluorescence emitted by the FRP, but must stop unwanted excitation light very effectively.

Light sources

The choice of a light source in fluorescence microscopy depends on the excitation spectrum of the FRP. Light sources such as xenon arcs and halogen–quartz lamps or mercury arcs which emit a more or less continuous spectrum with strong emission lines are chosen in relation to the absorption peak of the FRP. For quantitative microfluorometry these light sources should be operated from highly stabilised current supplies. With xenon and mercury arcs good results are obtained when using DC current supplies. The halogen–quartz lamp has been found to be sufficiently bright for some applications in quantitative fluorescence microscopy (Ploem, 1970*a*) and can be switched off and on easily and frequently without damage to the lamp.

For very low concentrations of a FRP, for a FRP which fluoresces with a relatively low quantum efficiency and for a FRP which requires ultraviolet or violet excitation light, the mercury 100-watt lamp (e.g. Osram HBO 100) or the xenon 75-watt lamp (e.g. Osram XBO 75) operated on DC power supply are suitable. Because of the relative stability of the arcs these lamps are also suitable for quantitative microfluorometry (Ploem, 1971). They should, however, be burnt-in under conditions of low mechanical vibration (e.g. during the night) and during this period the use of an AC voltage stabiliser in order to overcome large voltage fluctations of the electric mains, in addition to a highly stabilised DC supply, is recommended. This results in more stable burning points on the anode and cathode of the lamp and a more stable arc with consequent greater stability of light output for microfluorometry.

Lamp housings

A lamp housing should permit a good alignment of the light source and flexible positioning of the collecting lens. For efficient illumination it is of the utmost importance to choose a collecting lens in which the focused image of the filament or arc will fill the entire entrance pupil of the condenser used in transmitted illumination or the pupil of the objective used for epi-illumination. Under these conditions correct Köhler illumination will be obtained resulting in an even distribution of illumination of the microscopic field in the object plane (Ploem, 1973).

It is important to verify that the position of any reflecting mirrors and the collecting lens in the lamp housing does not change because of a rise in temperature during the period of measurements. In general it is advisable to have a warming-up period and then centre the collecting lens and mirrors again before starting the fluorescence readings. Uneven distribution of excitation light in the object plane, due for instance to incorrect focusing of the collecting lens causing an image of the filament or arc of the lamp in the object plane, may give rise to large variations in the fluorescence readings, depending on the location of the specimen in the microscopic field. Large and small objects having exactly the same amount of FRP may give different readings in that case. Evenness of illumination in the object plane can be tested with the 'nine position' test described by Boehm (1972) using several fluorescence readings taken from different locations on a fluorescent GG 21 Schott glass slide used as a test object.

Excitation filters

The three main types of filter used at the present time in fluorescence microscopy are coloured glass filters, band-interference filters, and short-wave pass-interference filters. These filters can also be roughly divided into wide-band filters and narrow-band filters. For the general purposes of fluorescence microscopy an excitation filter can be considered to be of the wide-band type if the half width (HW = the width of the main transmission band at 50 % of the peak transmission) is more than 50 nm and to be a narrow-band filter if the half width is less than 20 nm.

Most fluorochromes and fluorophores have rather broad main excitation bands. When it is necessary to obtain a strong fluorescence signal it is therefore rarely desirable to use an excitation filter with a veiy narrow main transmission band, for example having a half width of about 5 nm, even if this corresponds to the peak of the excitation curve of the FRP. The only exceptions are applicatons where a strong emission line of a high-pressure arc corresponds to the peak excitation of a FRP. In other cases a considerable part of the emission of the lamp which could be used for the excitation of the FRP is lost with the use of excitation filters having a very narrow half width. Wide-band excitation, on the other hand, is also often undesirable since together with the fluorescence of the FRP, autofluorescence of tissue components and optical parts is excited by the shorter wavelengths transmitted by such a filter. If two different FRPs are present in the microscopical specimen the use of a wide-band excitation filter is seldom permitted in microfluorometry since both may fluoresce simultaneously and lead to readings which are difficult to interpret. In most applications excitation filters with a half width of about 20 nm are satisfactory.

Recently short-wave pass-interference filters have been manufactured with a steep edge towards the longer wavelength side. Several of these filters have been described (Richards & Waters, 1967; Rygaard & Olson, 1969; Kraft, 1970; Tomlinson, 1970; Ploem, 1971). This type of filter does not only have a high transmittance (more than 80 %), but even more important, eliminates the unwanted excitation light towards the longer wavelength side very effectively, often to less than 10^{-5} of the original energy emitted by the lamp in this wavelength range (Fig. 1). Ploem (1971) has suggested the use of a conventional barrier filter type of glass acting as a long-wave pass (LP) filter in combination with the aforementioned short-wave pass (SP) interference filter. The long-wave pass filter is used in this case in the excitation light beam to absorb the shorter wavelengths that cause autofluorescence or excite another FRP in a double staining procedure and pass only the longer wavelengths wanted for the excitation of a specific FRP. Thus using combinations of LP and SP filters, many excitation filters can be tailored (Fig. 2) according to the requirements of a certain FRP (Ploem, 1975).

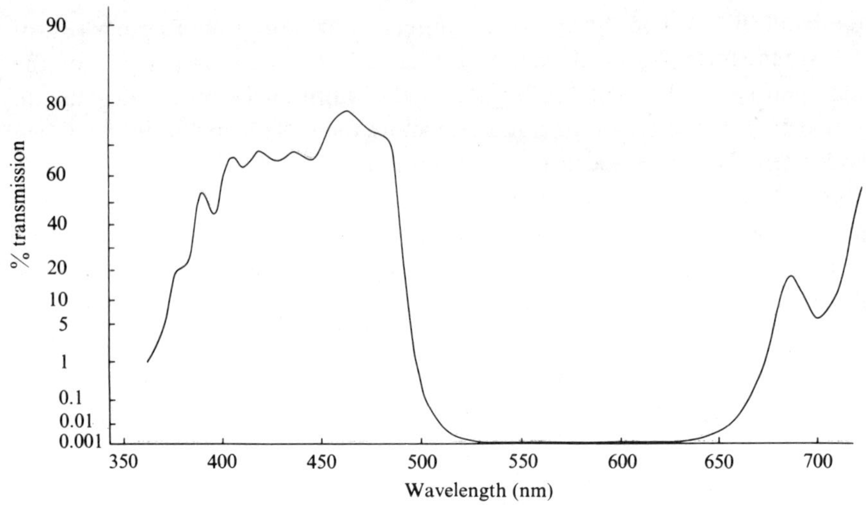

Fig. 1. Transmission curve of a typical 2 × KP 490* (SP 490) short wave pass-interference filter for selecting (transmitting) violet–blue excitation light and blocking (reflecting) unwanted green and orange excitation light.

* KP 490 (German) = SP 490 (English).

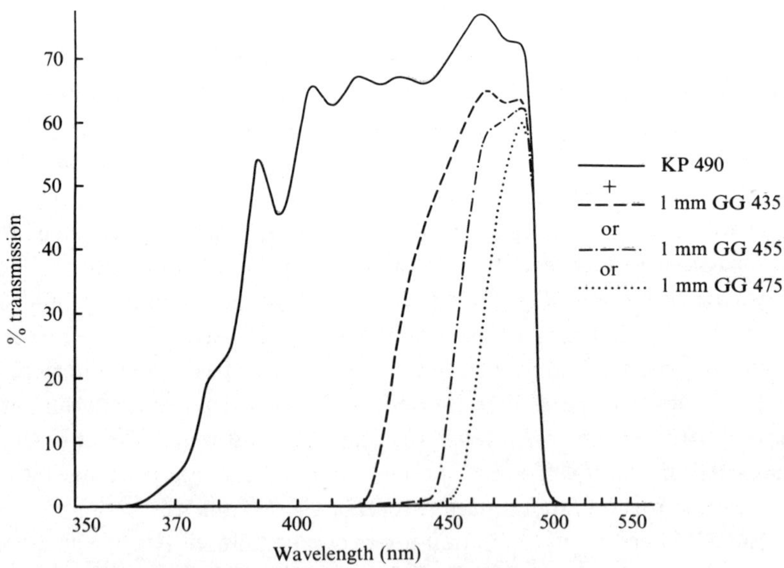

Fig. 2. Transmission curves of a KP 490* (SP 490) short-wave pass-interference filter in combination with several LP filters having cut-off values at 430, 450 and 470 nm respectively. The combined LP–SP filter results in excitation filters with a different bandwidth.

* KP 490 (German) = SP 490 (English).

Illuminating systems

The main types of illuminating systems used in fluorescence microscopy are transmitted dark-field and bright-field illumination, incident dark-field illumination with a ring condenser or incident illumination through an ordinary microscope objective for bright-field illumination. In routine fluorescence microscopy substage dark-field illumination or epi-illumination through an ordinary objective are generally used. In microfluorometry, however, combinations of the illumination types described above are often necessary.

Dark-field condensers. With the use of a dark-field condenser the cone of the excitation light, focused on the specimen, does not enter the microscope objective directly. A hollow cone of light, whose apex falls in the plane of the specimen, is formed. Only when a microscopical specimen is present can light be refracted and thus enter the objective, resulting in a dark-field image. Especially with the dark-field condensers used for fluorescence microscopy, which should have a relatively high numerical aperture (NA), very careful focusing and exact centring of the condenser are prerequisites for obtaining the reproducible illumination conditions required for constant fluorescence readings.

Bright-field condensers. For routine fluorescence microscopy bright-field condensers are rarely used since dark-field condensers give a much better image contrast. In microfluorometry they are, however, used for searching out and selecting suitable areas in the microscopical preparation, on the basis of their absorption and fluorescence image in transmitted illumination. In microfluorometry it is necessary to bring the object into a measuring field before measurement. Since searching and the careful centring of an object within a measuring diaphragm are often time-consuming, this cannot be done by using strong epi-illumination, since this leads to unwanted rapid fading of the FRP. However, several FRPs also have an absorption image which can be observed by transmitted bright-field illumination using a low-voltage lamp as a light source, and fading of the FRP is practically absent under these conditions. In some double-staining techniques one of the FRPs can only be found by fluorescence microscopy in the same microscopic specimen (Cornelisse & Ploem, 1976). In this case, the same substage bright-field condenser must also be used for fluorescence microscopy. With modern high-quality interference filters of the short-wave pass-interference type it is often possible to obtain a satisfactory image contrast with substage bright-field condensers.

Ring condensers. With a ring condenser of the Ultropak type (Leitz, Wetzlar) dark-field incident illumination is possible. With this incident light illuminator, a vertically adjustable ring condenser surrounds the objective concentrically and directs the light reflected by the ring condenser onto the object. Very long working distances are an outstanding feature of these objectives: when inserted into the microscopical specimen they permit the examination of deep layers within the specimen, which is impossible with ordinary microscope objectives.

Bright-field vertical illuminators. Multi-wavelength vertical illuminators are now a standard device on most microscope stands. They permit bright-field epi-illumination through a conventional objective in various wavelength bands. Dichromatic beam splitters, also known as dichroic mirrors, each coated for the reflection of a certain wavelength range can be brought into the incident light path (Ploem, 1967). A dichromatic beam splitter is placed at an angle of 45° to the optical axis and reflects exciting rays into the objective. The objective now functions also as a condenser and concentrates the excitation light on the specimen. Dichromatic beam splitters have a high reflectance for excitation light and a high transmittance for fluorescence light. The spectral characteristics of a dichromatic beam splitter designed for the reflection of green light and the transmission of orange–red light are shown in Fig. 3. From this it can be understood that blue–green excitation light is reflected, whereas unwanted orange–red and violet–ultraviolet excitation light is transmitted. The dichromatic beam splitter thus has the function of an excitation filter which removes unwanted excitation light from the illuminating beam (Fig. 4). The mirror has also the function of a barrier filter by eliminating most of the green exciting rays that are reflected from the glass surfaces of lenses of the objective and the cover glass. An extra barrier filter has, however, still to be used to absorb the small percentage of reflected green rays transmitted by the dichromatic beam splitter.

Objectives

For dark-field illumination, low to moderate power dry objectives are generally used. Oil-immersion objectives with an NA of more than 0.95 are in general not suited for dark-field illumination since these objectives would still collect light from the illuminating cone of the condenser, which would be visible as unwanted excitation light. In epi-illumination, however, the illumination of the specimen and the collection of fluorescence light is performed by the objective alone. Since both the illumination of the specimen and the collection of fluorescence light are proportional to the square of the

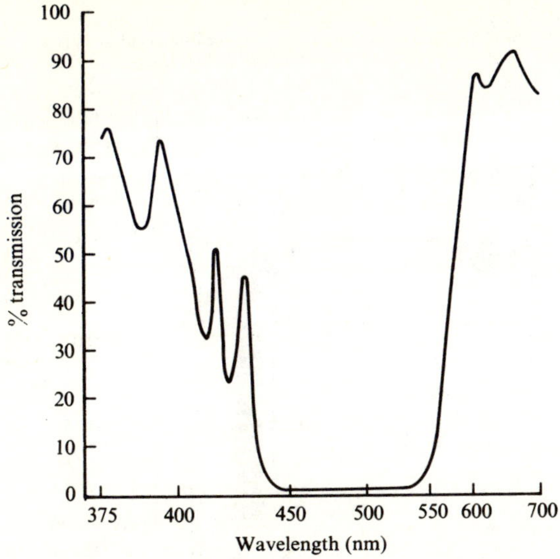

Fig. 3. Transmission curve of a dichromatic beam splitter (DS) reflecting green light and transmitting orange and red light. The reflectance is approximately the inverse of the transmittance and both are measured with the beam splitter placed at an angle of 45° to the optical axis.

NA of the objective, the efficiency of fluorescence microscopy using epi-illumination depends on the fourth power of the NA of the objective used. From this it will be clear that special objectives were developed for epi-illumination that had moderate magnification but maximal NA. For example an immersion objective 40×/NA 1.30 (Leitz, Wetzlar) and an immersion objective 63×/NA 1.40 (Zeiss, Oberkochen) were manufactured. Many more special objectives for epi-illumination are now available, also from other manufacturers.

For epi-illumination, immersion objectives are better suited than dry objectives, since the amount of incident light reflected from glass to air interfaces is much more than that from glass to oil, glycerine or water interfaces.

Eyepieces (*oculars*)

If fluorescent objects in the microscopic preparation have to be located and centred in the measuring diaphragm, the danger of fading is evident. To minimise this danger the intensity of the excitation light should be so low that the fluorescent objects are just visible and can be brought into the measuring field. Because the brightness per unit area of the microscope preparation

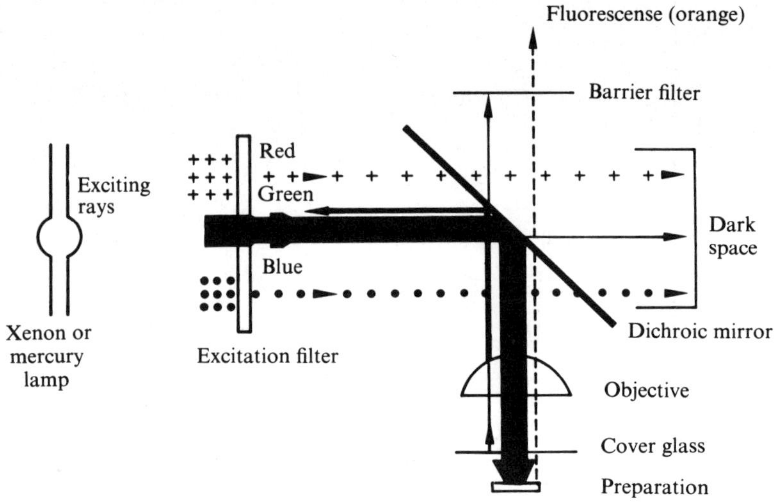

Fig. 4. Light path in a vertical illuminator equipped with a dichroic mirror (dichromatic beam splitter) placed at 45° to the main optical axis of the microscope. From a light source which has a high level of intrinsic brilliancy the exciting rays, mainly green light, are selected by an excitation filter. Some of the exciting rays, which are reflected back from the glass surfaces of the objective and the cover glass, are absorbed by a barrier filter. Only the orange fluorescence from a microscopical preparation can be transmitted by the barrier filter.

decreases with the square of the magnification of the eyepiece, the use of low-power eyepieces (e.g. 4–6.3 ×) is often indicated.

Since objectives with a high NA have a relatively large magnification factor, the use of low-power eyepieces in combination with such an objective is often indispensable for obtaining a moderate total magnification resulting in a high collecting power for fluorescence light.

It should be realised that very effective collecting of fluorescence light in microfluorometry is often better than the use of very strong excitation light, since the latter often causes rapid fading of the FRP.

Secondary (barrier) filters

To obtain optimal brightness and image contrast, appropriate barrier filters have to be selected in combination with each excitation filter combination. Barrier-type filters transmit fluorescence with wavelengths longer than the cut-off value of the filter. Narrow-band filters, which only transmit the peak fluorescence of a certain FRP, can also be used as a secondary filter. Hereafter the latter type of filter will be described as a fluorescence selection filter.

The cut-off value of the barrier filter should be chosen to prevent the transmission of unwanted excitation light. It is, however, inadvisable to apply secondary filter combinations chosen with a cut-off value at the safe (longer wavelengths) side, which would allow observation of only a part of the fluorescence spectrum (Van der Ploeg & Ploem, 1972). The resulting loss in fluorescence is undesirable in microfluorometry, since for each relative unit of fluorescence obtained, a certain percentage of fading of the FRP must be accepted. Often, therefore, optimal collecting of fluorescence light is of more importance in microfluorometry than very strong excitation of the FRP.

Recently a new type of fluorescence selection filter has been introduced composed of a barrier-type of coloured glass filter in combination with a short-wave pass-interference filter. Such fluorescence selection filters have a transmission curve with a steep edge both at the shorter and at the longer wavelength side. This filter combination has high transmittance for the fluorescence wavelengths and a strong suppression of unwanted excitation light and unwanted fluorescence light (L. Trapp, personal communication; Ploem, 1975).

In microfluorometry of double stained specimens this type of fluorescence selection filter is often a prerequisite for a sufficient separation of the fluorescence of two FRPs.

Microscope stands

Conventional small microscope stands are often equipped only for illumination with transmitted light. In Fig. 5 a simplified ray diagram of a simple microscope with a substage dark-field condenser is shown. These inexpensive stands can be used in combination with a relative simple microfluorometer as described by Taylor, Heimer & Lidwell (1971). For several specialised applications this type of instrument gives satisfactory results.

Relatively simple small microscope stands have recently become commercially available equipped for incident illumination (Fig. 6) using a vertical illuminator after Ploem (e.g. from Leitz, Wetzlar; Zeiss, Oberkochen; Reichert, Vienna; Vickers, York; American Optical, Buffalo; Olympus, Tokyo; Zeiss, Jena; Polskie Zaklady Optyczne, Warsaw).

For general purpose applications greater flexibility can be achieved with a stand permitting both substage as well as incident illumination (Fig. 7).

Microfluorometry can be best performed with epi-illumination (Rigler, 1966; Ruch, 1970; Boehm, 1972; Rost, 1972), since the functions of illuminating and observation of the specimen are combined in one objective. The excitation conditions are therefore reproducible during movement of the specimen with the microscope stage as long as the focusing is maintained.

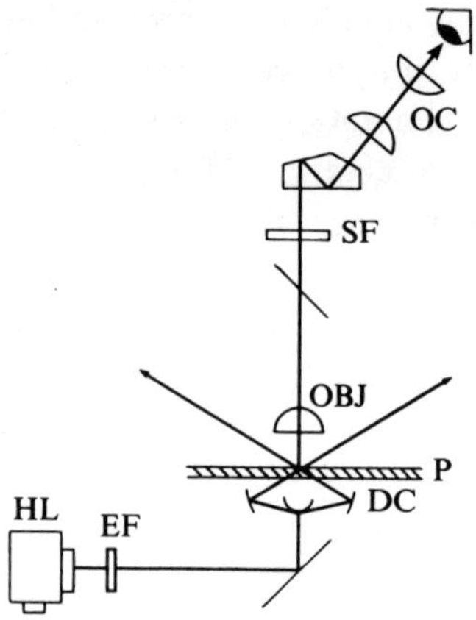

Fig. 5. Schematic representation of the light path in a fluorescence microscope equipped with dark-field illumination. A light source (HL) and an excitation filter (EF) provide the excitation light which is concentrated with a dark-field condenser (DC) on the microscopical preparation P. A secondary filter (SF) absorbs unwanted excitation light diffracted by the specimen, and transmits fluorescence light toward the eyepieces, OC.

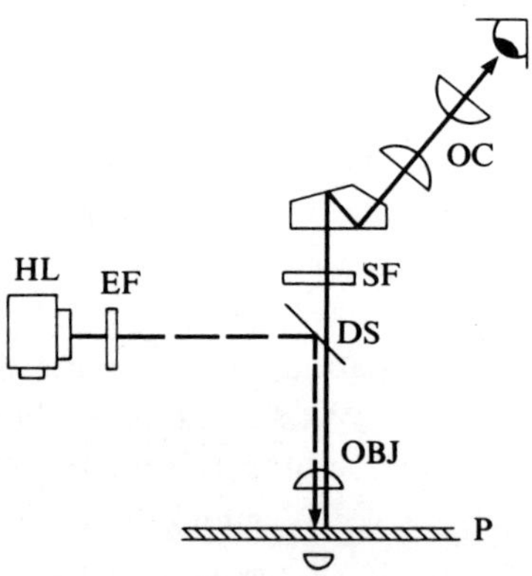

Fig. 6. Schematic representation of a fluorescence microscope equipped for epi-illumination. A dichromatic beam splitter (DS) deflects the excitation light onto a microscopical preparation and transmits the fluorescence light towards the eyepieces. The optical parts are indicated with the same symbols as in Fig. 5.

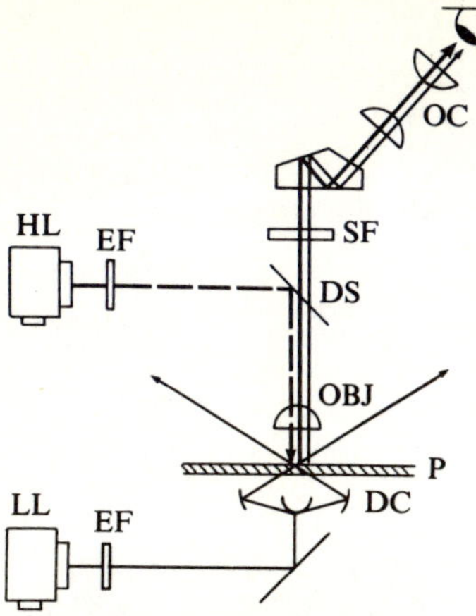

Fig. 7. Schematic representation of a fluorescence microscope equipped for simultaneous illumination with transmitted and incident light. LL, low light intensity illumination; HL, high light intensity illumination; other symbols as in Fig. 5.

The exact alignment of a substage condenser with the objective is thus avoided and the substage condenser can now be used as a second illuminating system for searching for objects to be measured.

Moreover, with incident illumination the top layer of the stained specimen, which is directly exposed to the exciting radiation, is turned towards the photomultiplier, so the emitted fluorescence light does not have to pass through the total vertical dimension of the specimen before it can be measured. This is of importance if measurements have to be performed in a spectral range where partial reabsorption of the fluorescence light by the FRP must be considered.

Specialised microscope stands for microfluorometry make use of combined transmitted illumination for locating the objects to be measured on the microscope slide and epi-illumination for measurement. In Fig. 8 a simplified diagram of an example of such an instrument is depicted. Two light sources are available. Transmitted illumination is obtained from a tungsten lamp or a halogen–quartz lamp providing illumination with low light intensity (LL) which prevents fading of the FRP during centring of objects in the microscopical field. A bright-field condenser is used for visualising objects on the basis of their absorption. The field diaphragm (FD_1) is kept fully opened to illuminate the whole field while searching for objects to be measured. Epi-illumination is obtained from an HBO 100 mercury high-pressure arc operated on a highly stabilised DC power supply, using a vertical illuminator for concentration of the incident light on the specimen. The desired wavelength range is selected with a combination of excitation filters (EF_2). The

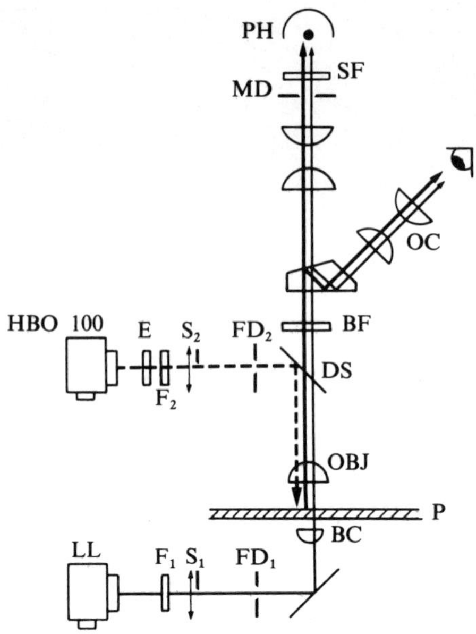

Fig. 8. Schematic representation of a microfluorometer equipped with electronically controlled shutters S_1 and S_2 in the transmitted and incident light path respectively. A halogen–quartz or tungsten lamp is used for low intensity level (LL) transmitted illumination. A filter F_1 is used to select the wavelength band with which the FRP gives a visible absorption image using a bright-field condenser (BC). The substage field diaphragm (FD_1) is opened to illuminate the entire microscopic field. An HBO 100 high-pressure mercury arc is used for epi-illumination. With the excitation filter combination EF_2, narrow-band excitation is selected. The field diaphragm FD_2 of the vertical illuminator is opened so as to illuminate an area just larger than that of the measuring diaphragm (MD). In addition to a barrier filter (BF) to block unwanted excitation light, another secondary filter (SF) just below the photomultiplier (PH) is used to select the wavelength range to be measured.

field diaphragm (FD_2) of the vertical illuminator is of such a size that it is only slightly larger than the measuring diaphragm. With some modern fluorometers the measuring field is visible as an illuminated spot superimposed on the image of the microscopical specimen (e.g. MPV II, Leitz; MP 03, Zeiss).

Electrical shutters are mounted in both the transmitted and incident light path (S_1 and S_2) and they are operated automatically by an electronic system (Ruch & Leeman, 1973; Deelder & Ploem, 1974; Wasmund, 1976). During the period of locating the objects to be measured, the substage electrical shutter is open. When the objects to be measured have been centred in the measuring field the measuring operation is started by pressing a button which initiates

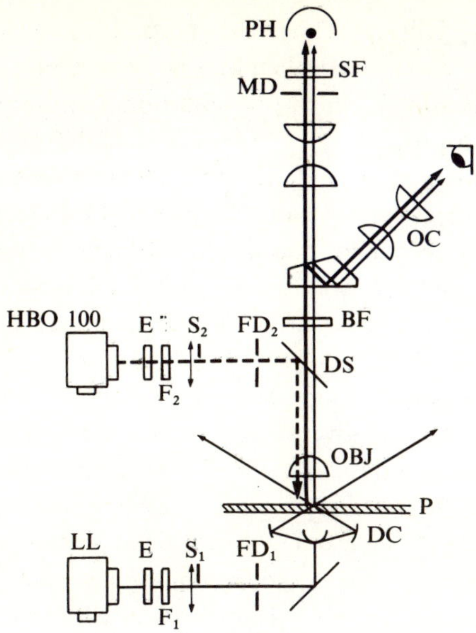

Fig. 9. Schematic representation of a microfluorometer using transmitted dark-field (DC) illumination with a low light intensity (e.g. from a halogen–quartz lamp) for finding the fluorescent objects to be measured and epi-illumination for measurement of fluorescence. The symbols used are the same as in Fig. 8.

the electronic circuit. The shutter in the transmitted light path is now closed and the shutter in the incident light path opened for a predetermined period which can be varied between about 0.01 and 1 s. Then the shutter in the incident light path is closed again and the shutter in the transmitted light path is opened automatically for the continuation of the search for objects to be measured until the next measuring sequence is started. This procedure allows very short exposure times for the measurement of a rapidly fading FRP, especially with weakly fluorescing objects requiring relatively strong excitation light. With this type of instrument a separate viewing facility for focusing and centring the objects within the measuring field is unnecessary since searching for objects and centring them in the measuring field is carried out exclusively through the binocular head. This facility and the automatic operation of the shutters described above permits easy and rapid measurement of a large number of objects in routine work.

For special applications other combinations of illuminating systems are required. Fig. 9 is a schematic diagram of a microfluorometer used for measurement of a FRP which does not give a visible absorption image. The specimen can only be visualised by fluorescence, as is the case for instance with fluorophores obtained from certain neurotransmitters after treatment with formaldehyde vapour (Falck, Hillarp, Thieme & Torp, 1962; Ploem, 1971). Searching for objects to be measured can then best be carried out with dark-field illumination using a light source which provides illumination with low light intensity (LL) in order to prevent rapid fading of the FRP during centring of the objects within the measuring field. Since the fluorescence thus obtained will be very weak, an optical system with maximal light collecting power is necessary for observing the specimen. The aforementioned combination of moderate-power objectives (40×/NA 1.30, 63×/NA 1.40) used with low-power eyepieces (6.3×, 4×) is required. The automated measuring procedure is the same as described above.

In other applications, where for example double staining is used, one of the final reaction products of the staining may give an absorption and a fluorescent image whereas the other reaction product may be detectable only by its fluorescence. Cornelisse & Ploem (1976) described such a staining procedure for application in exfoliative cytology. Nuclear DNA is stained with acriflavine–Feulgen, resulting in an absorption image used for finding the nuclei on the slide. The DNA content of the nuclei is subsequently measured by microfluorometry of the acriflavine fluorescence of the nuclei using incident illumination (Fig. 10). The cytoplasm of the cells is visualised by a blue fluorescence of the protein stained with isothiocyanatostilbene sulphonic acid using bright-field illumination with a substage condenser. This fluorescence is not measured but used only for recognising the cell types.

Microfluorometers

Since the author has extensive experience with microfluorometers of the type MPV II from Leitz and the MP 03 from Zeiss, the above descriptions of microscopes suited for microfluorometry refer particularly to these instruments. Many other microfluorometers, however, are commercially available or described in the literature, which have simpler, similar or more complex functions.

Relatively simple and inexpensive instruments may give good results in special applications. Taylor *et al.* (1971) described a fibre optic probe which forms part of the microscope eyepiece. Only the tip of the probe is sensitive to light. By manipulating the stage of the microscope a fluorescent particle can be brought under the tip of the probe and is measured. A photomultiplier is connected to the end of the probe. This special eyepiece can be adapted to most conventional fluorescence microscopes.

Somewhat more complex photometer accessories like the MPV I from Leitz and the MP 01 from Zeiss can be adapted to most research microscope stands of Leitz and Zeiss. This type of photometer requires, however, a

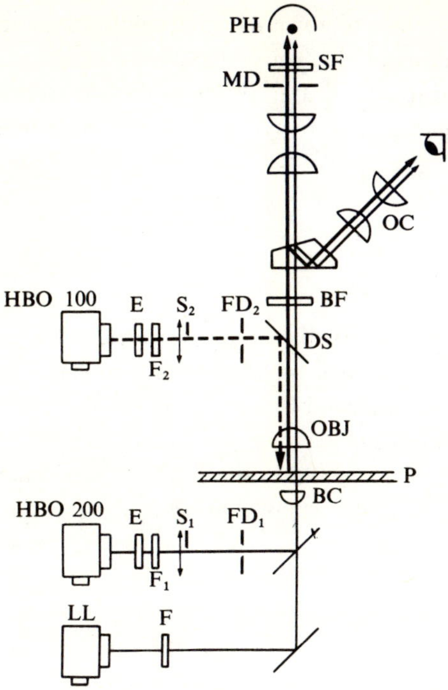

Fig. 10. Schematic representation of a microfluorometer for the examination of double stained specimens using transmitted bright-field (BC) illumination of low light (LL) intensity. Searching for objects is performed on the basis of their absorption image. Subsequently high-intensity illumination from an HBO 200 high-pressure mercury arc is used to locate other structures on the basis of the fluorescence, also using substage bright-fleld illumination with a high-quality excitation filter set (EF_1) to obtain an acceptable image contrast. Microfluorometry is performed with epi-illumination of a small field (FD_2). The symbols used are the same as in Fig. 8.

separate eyepiece for viewing the object simultaneously with an image of the measuring diaphragm. Measuring large numbers of objects is more tiring than with the more sophisticated MPV II and MP 03 models previously discussed, which permit the visualisation of the image of the measuring diaphragm in the binocular head. Recently multi-purpose large research microscope stands with infinity optics have become available (Univar: Reichert, Vienna; Axiomat: Zeiss, Oberkochen). These stands adapt simultaneously more than two lamp housings for transmitted and incident illumination, an automatic camera and a photometer. Easy photographic documentation of cells to be measured is thus possible.

The retardation rings for phase contrast are not fitted permanently into the objectives but can be inserted when required into the pupil image. This method offers special advantages for microfluorometry with epi-illumination of a FRP in, for instance, vitally stained living cells which are best visualised with phase contrast microscopy using transmitted illumination.

The reader should consult the manufacturers about the many possibilities of these instruments.

Methodology of microfluorometry

For the original papers and review articles discussing the methodology of microfluorometry the reader should consult Ruch & Bosshard (1963), Ruch (1964, 1970), Rost (1972) and Prenna, Mazzini & Cova (1974 *b*).

Some of the advantages of microfluorometry over microabsorptiometry in special applications have already been discussed in the introduction, such as the lack of distributional error in microfluorometry, owing to the fact that at high levels of transmittance (see below) overlap of chromophores becomes improbable.

Microabsorptiometry suffers from distributional error (see Goldstein, this volume, when the absorbing substance is not distributed homogeneously in the object to be measured, which is often the case in biological specimens. This distributional error is due to the relationship between transmittance (T) and the absorbance.

The absorbance (A) is defined by the equation $A = -\log_{10}T$. The transmittance (T) is determined by measuring the light transmitted through the object (I) with a microphotometer and comparing it with the light transmitted in an empty spot in the microscopical preparation (I_0) outside the object to be measured. The transmittance then can be calculated from $T = I/I_0$.

An example is given in Fig. 11 in which a nucleus A with homogeneously distributed chromatin is compared with a nucleus B in which all the chromatin is present in several dense clusters occupying only 10 % of the area of the nucleus. In both these hypothetical nuclei the amount of chromatin is exactly the same. Owing to the distributional error the apparent chromatin contents (based on the absorbance calculated from the transmittance measured by adapting a measuring diaphragm of a microphotometer to fit around the two nuclei), differ by a factor of more than 3. In modern instruments this error can be avoided by scanning microabsorptiometry using point-to-point measurements. If the scanning spot is sufficiently small compared with the size of the chromatin clusters, then these can be considered as homogeneous in relation to the size (e.g. 0.25–0.5 μm) of the measuring spot. The integration of a few hundred measured local absorbances results in the determination

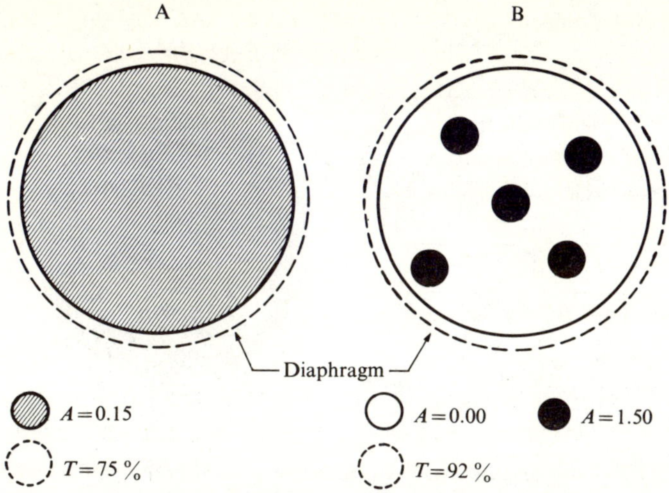

Fig. 11. A hypothetical nucleus A with homogeneously distributed chromatin with an absorbance of 0.15, and a nucleus B in which the same amount of chromatin is present in several clusters occupying only 10 % of the area of the nucleus and having a local absorbance of 1.50. The remainder of nucleus B has an absorbance of 0.00. The total transmittance (T) is measured within an area (size of measuring diaphragm) somewhat larger than the nucleus.

of the total chromatin content of the nucleus, avoiding distributional error. This method, however, requires a specialised instrument (see Goldstein, this volume, which can scan the entire area of a nucleus (Fig. 12) and integrate the local absorbances using analogue electronic methods (scanning and integrating microdensitometer GN 5: Barr & Stroud, Glasgow; or M 85: Vickers, York) or digital methods using a stepping stage under the control of a small digital computer (SMP 01: Zeiss, Oberkochen; or MPV II: Leitz, Wetzlar).

In microfluorometry, however, the total amount of substance can be obtained quite simply. If the intensity of the emitted fluorescence is proportional to the amount of chromatin present (this depends on the results of the cytochemical staining reaction used) the fluorescence of the entire nucleus can be collected by bringing the image of the nucleus to coincide with only one large measuring diaphragm (Fig. 12), and thus measuring it in one operation. Quite simple microfluorometers can be used for this procedure (Taylor *et al.*, 1971).

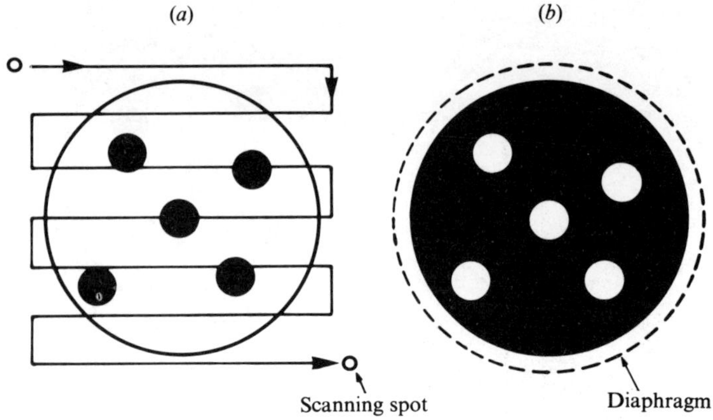

Fig. 12. Scanning microabsorptiometry and microfluorometry. (*a*) Determination of the total amount of inhomogeneously distributed chromatin in a nucleus by point-to-point scanning with a spot of e.g. 0.5 μm^2. This spot covers in a raster the entire nucleus and the few hundred optical densities so measured are integrated to obtain the total chromatin content of the nucleus. When the measuring spot is sufficiently small compared with the size of the chromatin clusters these can be considered as homogeneous and the distributional error will be relatively low. (*b*) Determination of the total amount of chromatin by fluorescence. It is assumed that the intensity of the emitted fluorescence in each part of the nucleus is proportional to the amount of chromatin present (this depends on the type of cytochemical staining). The image of the entire nucleus is brought within one large measuring diaphragm and the fluorescence can be measured in one step. The distributional error will now be negligible.

Signal-to-noise ratio

Stained specimens which have a high level of transmittance can seldom be measured by microdensitometry. The sensitivity of microfluorometry, however, often enables a relatively accurate measurement with preparations which show a very low absorption provided a FRP is present which fluoresces with sufficient quantum efficiency. Adequate measurements under these conditions are, however, only possible if the signal-to-noise ratio of the system is favourable. The fluorescence signal must be strong enough (i.e. there must be sufficient photons to reduce noise) and the image contrast must be optimal. This can only be achieved by using a light source with a high intrinsic brilliancy in combination with a high-quality primary and secondary filter system which reduces unwanted excitation light in the background and which results in a strong fluorescence. For this reason an extensive section dealing with light sources and filters has been included in this reveiw. For this same reason the methods available for efficient collection of fluorescent light with objectives of high NA are discussed.

It is evident that with relatively weakly fluorescing objects the light measured in the background must be quite low. Unwanted excitation light levels must in that case sometimes be reduced to less than 10^{-7} of the original intensity in the wavelength range where the fluorescence of the FRP is measured.

It is advisable to test the filter combination used in microfluorometry with standard microscopical preparations embedded in a mounting medium which has a low level of autofluorescence. Microcapillaries filled with a fluorochrome solution of a certain concentration (Sernetz & Thaer, 1970; Jongsma, Hijmans & Ploem, 1971) or spherical Sepharose beads of microscopical dimensions, stained with a certain fluorochrome (Haaijman & Van Dalen, 1974) can be used for this purpose.

One should always measure (1) the dark current of the photomultiplier, (2) the background light in an area outside the fluorescent objects, and (3) the signal obtained from a standard fluorescent object. In general one requires, even with fluorescent signals of low intensity, a signal-to-noise ratio of better than 20:1 using a measuring period of about 1/100th of a second (available for integrating photons).

When autofluorescence forms a considerable part of the background signal, it should be noted that autofluorescent material in biological specimens often has very wide excitation and emission spectra due to the many chemical components of such a material. With wide-band excitation and the conventional barrier-type of secondary filter which transmits a wide spectral range, it is almost impossible to avoid measuring some autofluorescence in the background outside the specifically stained objects. Fortunately, most specific FRPs have narrower excitation and fluorescence peaks. In order to obtain selective excitation and observation of the FRP one should excite it at the peak of its excitation curve and observe its fluorescence at the peak of its emission curve. The first can be achieved by narrow-band excitation with a combination of LP+SP excitation filters and the second with LP+SP secondary filter systems for fluorescence selection as already discussed in the section on filter systems. In the measuring instrument as well as a barrier filter an extra secondary filter (SF) system used as a fluorescence selection filter is inserted in a filter slot above the binocular tube (SF in Figs. 8, 9 and 10). This filter transmits of course only a limited selection-band from the total fluorescence image. The resulting image will be monochromatic. The entire polychromatic fluorescence image and the image contrast remain, however, visible in the binocular head. By using low-power (e.g. 4×, 6.3×) eyepieces to obtain a bright image, unwanted excitation light can be detected if present and autofluorescence can be better observed, for the human eye is much better able to judge differences in colour in the bright images thus obtained. It is important to note that it is difficult to judge the signal-to-noise ratio in an image of apparently normal intensity.

Fluorescence standard and calibration of the microfluorometer

In microdensitometry the intensity of the light transmitted through microscopical objects is compared with the transmission through an empty part of the microscopical preparation (blank). Changes in the light emission of the lamp will influence the readings from the stained object and the blank in a proportional way, so microdensitometry is therefore independent of such changes. In microfluorometry the situation is different. The reading in an empty part of the microscopic preparation is the result of the sum of the measured dark current of the photomultiplier, the photocurrent still caused by unwanted excitation light and autofluorescence, so this signal cannot be used as a reference signal. Moreover the signal from the specifically stained FRP changes in a non-linear way with changes in the high voltage supply and the electronic amplification. The only way to calibrate a microfluorometer is by using a standard fluorescent preparation. A non-fading slide of uranyl glass (GG 21 from Schott, Mainz) has been proposed for this purpose (Rigler, 1966).

When epi-illumination is used, quite reproducible readings can be obtained with an oil-immersion objective just touching a 2–4 mm thick uranyl glass slide (Ploem, 1970*b*; Jongsma *et al.*, 1971). Although this standard is simple to use, and is relatively insensitive to fluctuations in temperature, it can only be used for calibrating a microfluorometer for relatively short periods of time. The reason for this is that only during such short periods can changes in the position of the collecting lens in the lamp housing and the aperture of the field diaphragm of the vertical illuminator be avoided. Any changes in these parts would result in a different reading from the uranyl glass due to the thickness of the slide (Jongsma *et al.*, 1971). This uranyl glass standard can be successfully used during a measuring period of several hours to compensate for changes in the readings caused by a change in the temperature of the lamp, filters or the electronic equipment and photomultiplier.

For a discussion of different kinds of standards in microfluorometry the reader should consult a review by Ploem (1975).

Fading of fluorescence (photodecomposition)

Fading of fluorescence is defined here as a decrease in fluorescence during excitation of a FRP, regardless of its exact cause. The degree of fading depends on the intensity of the excitation light and the exposure time. The rate of photodecomposition depends, however, mainly on the absorbed intensity of excitation light. With many FRPs the fading is much more pronounced at higher intensities of excitation light (Jongsma *et al.*, 1971). As a consequence either the exposure time must be kept short or the intensity of the exciting illumination must be kept low in quantitative microfluorometry. In several applications of the method the photodecomposition rate may be high enough to produce a considerable decrease in fluorescence over time intervals as short as one second. We have already discussed the methods used to keep fading of the FRP to a minimum when using modern microfluorometers. Searching for objects to be measured is preferably done on the basis of an absorption image, or if the FRP has no visible absorption, on the basis of low-intensity transmitted excitation light. A light collecting system, consisting of a moderate power objective of high NA and an eyepiece with a low magnification must be used to collect as much as possible of the fluorescence for visual recognition of the specimen. Exposure times for measurement wih epi-illumination are generally kept short with the aid of electronic shutters in order to reduce fading as far as possible.

Quenching of fluorescence

Quenching is due to modification in the excited states of the fluorophore molecules. It results in a reduction of the fluorescence quantum yield and thus in a diminished fluorescence emission. When different concentrations of a FRP are obtained in one microscopical preparation, or the FRP is present in a different physicochemical environment, one should always give attention to the possibility of quenching. Quenching is a complex phenomenon and dependent on, among other factors, the concentration of the fluorophore, temperature, viscosity, and reversible or irreversible reactions of molecules in the excited state.

Reabsorption

Following an overlapping of the absorption and emission spectra of a fluorophore, reabsorption of the fluorescence light by the fluorophore itself may occur (Boehm & Sprenger, 1968; Prenna *et al.*, 1974*a*, *b*). As the absorption increases reabsorption affects a progressively larger area of the fluorescence spectrum. Then the emission intensities corresponding to the overlapping zone become progressively attenuated, and the spectrum of the emerging emission will be modified.

The phenomenon of reabsorption can be counteracted by measuring the fluorescence in a wavelength range where the excitation and emission curves are not overlapping. This can be done by using a barrier filter with a cut-off value at a longer wavelength. For instance in measuring the fluorescence of a conventional pararosaniline–Feulgen reaction where the mean local absorbance of the nuclei does not exceed 0.2, a barrier filter with a cut-off at 630 nm can be used to measure fluorescence outside the wavelength range where reabsorption disturbs the measurements (Boehm & Sprenger, 1968). Unfortunately many photomultipliers do not have a uniform spectral sensitivity over the green–red wavelength range in which many important fluorophores have their main emission. With many types of photomultiplier used in microfluorometry spectral sensitivity decreases quite rapidly as wavelength increases. In this case the spectral components that are unaffected by the reabsorption (longer wavelengths) are detected with a lower efficiency, and the reabsorbed light represents a greater percentage of the emitted light.

For a detailed discussion of this type of phenomenon the reader should consult Prenna *et al.* (1974*a*, *b*).

The relationship between fluorescence intensity and the concentration of the fluorophore

It has already been explained that microfluorometry makes it possible in principle to obtain a direct measurement without employing microdensitometric scanning techniques. This is, however, only true under the assumption that the FRPs in the various parts of the specimen contribute to the total fluorescence with quantities proportional to their concentration.

The relationship between fluorescence intensity (F) and the concentration (C) of a fluorophore (FRP) under conditions of epi-illumination is given in the following equation (Boehm & Sprenger, 1968):

$$F = I_0 Q_{\mathrm{rel}} (1 - \mathrm{e}^{-kcd}),$$

where F is the fluorescence emission,
I_0 is the incident light,
Q_{rel} is the relative quantum efficiency,
k is the molar extinction coefficient of the FRP,
c is the concentration of the fluorescent substance
and d is the thickness of the object.

If the relationship between fluorescence and the concentration of the FRP is calculated, a curve is obtained which approaches asymptotically an upper limit. At higher concentrations the fluorescence is thus no longer proportional to the concentration of the FRP. Only at lower concentrations is an approximately linear relation between fluorescence and concentration of a FRP found. With low local absorbances (0.1–0.2) the latter condition will obtain. In practice, however, a perfect linear relationship between fluorescence and concentration is not always required in order that quite reasonable measurements can be obtained in many applications with average absorbances of up to 0.4. Various local absorbances within a single nucleus (e.g. 0.1–0.6) do not give rise to a very large error, since the underestimation of fluorescence at high absorbances is compensated for by the measurement at lower absorbances. These conditions are, however, only met in a rather homogeneous cell population and not when very large faint nuclei are compared with very small dense ones.

Standard preparations

In contrast to readings in absorptiometry, microfluorometric readings are relative values. If, for instance, the electronic amplification of the photocurrent is increased then the readings also increase. The background value measured cannot be used to calibrate the fluorescence signal, since this reading depends mainly on autofluorescence and unwanted excitation light. The fluorescence readings obtained from one microscopical preparation (slide) can be compared only with each other. For instance, in measuring DNA values from cancer cells it is necessary to establish in each microscopical preparation the diploid human value from standard cells such as human leucocytes. If necessary such cells must be added to the preparation. It is also possible to use a standard of microscopical dimensions containing or stained with the same fluorophore as the specimens to be measured. If the amount of fluorophore is exactly known, as in the case of a defined amount of fluorochrome solution in a microcapillary (Sernetz & Thaer, 1970; Jongsma *et al.*, 1971), or a fluorophore of defined concentration bound to a Sepharose bead of known dimensions (Haaijman & Van Dalen, 1974), such standards can also be used as reference values.

Applications

The use of quantitative fluorescence microscopy has increased considerably in the last few years. In this review only the main applications will be briefly mentioned and the reader should consult the literature mentioned below for a more detailed discussion of the topic.

Microfluorometric determination of nucleic acids

A relatively large number of staining reactions have been described for obtaining fluorescence from structures containing nucleic acid. Only a few examples will be briefly described to illustrate this application.

For some original and review articles the reader should consult: Bosshard (1964), Rigler (1966), Ruch (1966, 1970), Boehm & Sprenger (1968), Nitsch, Murken & Brueck (1970), Pearse (1972), Boehm (1972), Haag (1973), Prenna, Leiva & Mazzini (1974*a*), Gill & Jotz (1974) and Steinkamp & Crissman (1974).

Acridine orange. This is probably one of the best known fluorochromes for staining nucleic acids. From the vast literature one can conclude that the exact binding mechanism is still under investigation. For reviews the reader should consult Pearse (1972) and Boehm (1972). Basically two different reactions must be considered: (1) intercalation of the acridine orange molecules within the double helical structure of DNA between the neighbouring base pairs of the two polynucleotide chains and (2) electrostatic or polar-type binding between the basic dye and the acidic phosphate groups of the polynucleotide chains. The staining procedures described in the literature are complex and a careful evaluation of many factors involved in the staining procedure is necessary for the interpretation of results (Boehm, 1972).

Acridine orange can be excited with blue excitation light (LP 450–SP 490). As a very much simplified staining result one can state that acridine orange produces green fluorescence from DNA and red fluorescence from RNA and denatured DNA (Rigler, 1966).

For a review article on cells stained supravitally with acridine orange the reader should consult West (1969).

Ethidium bromide. This dye stains DNA and RNA. It intercalates into DNA with an enhancement of its quantum yield. The dye has been extensively used in flow-through cytofluorometry after treatment of cells with RNAse (Dittrich & Göhde, 1969). The dye can be excited with blue and green excitation light and fluoresces red. For a review the reader should consult Le Pecq (1973).

Pararosaniline–Feulgen. This staining has been used for more than forty years as an absorption stain. With the introduction of green excitation light into fluorescence microscopy a sufficiently strong red fluorescence could be obtained from the FRP (Ploem, 1967). In the concentrations used for fluorescence microscopy the stained nucleus can be visualised both on the basis of its absorption (at about 560 nm) and its red fluorescence using excitation with the green mercury line at 546 nm (using an LP 515–SP 560 short-wave pass-interference filter). Because of the considerable overlap of the absorption and emission spectra and the relatively high absorbances found under standard staining conditions, attention should be given to the reabsorption already described.

Acriflavine–'Feulgen'. This is not considered by Gill & Jotz (1974) as a classical Feulgen stain since strong nucleus-specific acriflavine staining takes place also in the absence of sulphur dioxide. Due to the strong yellow–orange fluorescence obtained, matching well with the spectral sensitivity of many light-sensing elements (photomultipliers, vidicons, plumbicons), the staining is widely used in microfluorometry. With the usual staining procedures both an absorption image and a fluorescence image are obtained so that one has to avoid an inner filter effect by using barrier filters with a cut-off at longer wavelengths (about 590 nm).

An acriflavine–'Feulgen' staining for DNA can be combined with other fluorescent stains like for instance isothiocyanatostilbene sulphonic acid for protein staining (Cornelisse & Ploem, 1976). Using a two-wavelength vertical illuminator, a sequential visualisation of DNA and protein fluorescence is possible.

Quantitative immunofluorescence microscopy

The introduction by Coons, Creech & Jones (1941) of fluorescent protein tracers in immunological research has given an enormous stimulus to the application of fluorescence microscopy.

Since immunofluorescence is so widely used now, it is not surprising that quantitative methods are increasingly being used in this area. Pioneer work in immunomicrofluorometry has been carried out by Goldman (1960, 1967), Mansberg & Kusnetz (1966) and Nairn (1968).

The reader should consult original papers and review articles on this subject by Nairn, Herzog, Ward & de Boer (1969), Irvine, Chan & Williamson (1969), Ploem (1970*a*, *b*), Kaufman, Nester & Wasserman (1971), Levin *et al.* (1971), Taylor *et al.* (1971), Jongsma *et al.* (1971) and Taylor & Heimer (1975). These investigators have used conventional light sources like high-pressure mercury and xenon lamps or halogen–quartz lamps for their quantitative studies.

A general conclusion from this type of study may be that in principle immunofluorescence can be evaluated quantitatively in many applications. The methodology is complex and sufficient standardisation of substrates and antisera is a prerequisite. In a number of applications the observations made with visual evaluation were confirmed. Often the scale along which the fluorescence intensity could be evaluated was much longer than that used in the semi-quantitative visual evaluation, which is generally expressed from negative to four plusses. It appears that very bright fluorescence cannot be judged visually as well as it can with microfluorometry. In a number of studies data could be obtained which could not have been assessed by visual work, often because visual evaluation requires as a minimum twofold differences between fluorescence intensity, whereas in microfluorometry much smaller differences can be objectively registered (Cherry, 1970; Goldman, 1971; Killander *et al.*, 1973; Kolk, Samuel & Ruemke, 1974; Wachsmuth & Woodhams, 1974).

Immunofluorescence of small objects can be very weak and the availability of strong blue light from a laser has stimulated quantitative fluorescence microscopy using such light sources (Kaufman *et al.*, 1971; Bergquist & Nilsson, 1975; Wick, Schavenstein, Herzog & Steinbatz, 1975). The results are promising and an increased use of excitation by laser light can certainly be expected in the future. In immunological research the sorting of living cells with the help of membrane markers has become an indispensable technique. Loken & Herzenberger (1975) have described a fluorescence-activated cell sorter using fluorescent membrane markers to sort visible cells. This instrument also makes use of a laser as a light source. Developments along this line will certainly have a pronounced influence on the scope of applications in immunomicrofluorometry.

Microfluorometry of neurotransmitter fluorophores

Biogenic amines such as catecholamines and 5-hydroxytryptamine can be converted to highly fluorescent products by treating dried tissues with formaldehyde vapour (Falck *et al.*, 1962; Eränkö, 1964). For characterisation and identification of neurotransmitter fluorophores a microspectrofluorometer is used. This type of instrument for quantitative fluorescence microscopy has, in addition to the microfluorometers already described, special facilities for continuously varying the excitation wavelength with a monochromator and analysing the fluorescence emission with a second monochromator (Björklund, Ehinger & Falck, 1968; Rost & Pearse, 1971; Ploem, de Sterke, Bonnet & Wasmund, 1974). Microspectrofluorometry has been the main type of quantitative fluorescence microscopy used in neurotransmitter research. Conventional microfluorometry using the instruments described in this review is, however, used to quantitate formaldehyde-induced neurotransmitter fluorescence in cells and tissue sections (Lichtensteiger, 1970; Dowson & Laszlo, 1971; Jonsson, 1971; Tilders, Ploem & Smelik, 1974).

As with DNA microfluorometry the cells or selected parts in the tissue section to be measured are brought with the stage within the image of the measuring diaphragms of the microfluorometer. Since no absorption image is available in this application, searching for structures to be measured has to be done with substage dark-field illumination. Since neurotransmitter fluorophores fade rapidly under exposure to strong violet excitation light, only very weak excitation light can be used. Ploem *et al.* (1974) were able to use a halogen–quartz lamp for this purpose because the fluorescent light was collected efficiently with a moderate-power objective of high NA (40 × /NA 1.30) and a low-power eyepiece (4 ×).

Microfluorometry in kinetic studies

For kinetic studies in living tissues and cells or unfixed sections, microfluorometry is particularly suitable because of the high sensitivity that can be achieved. Deleterious effects on the specimen can be kept to a minimum by very effective collection of the fluorescence light using the least possible irradiation. The necessary methodology for this purpose has already been described elsewhere in this review.

In this type of application incident illumination with a ring condenser of the Ultropak type (Leitz, Wetzlar) is often used since with this system the objective can be pressed into the tissue. Many studies reported in the literature deal with oxidative enzyme systems in living cells which are analysed by measuring changes in the fluorescence of reduced pyridine nucleotide (NADH) bound to protein (Chance, 1962).

For descriptions of the methodology the reader should consult Kohen, Kohen, Thorell & Wagener (1973) and Kohen, Thorell, Kohen & Salmon (1974). In other applications cells are studied in a medium containing certain non-polar fluorescein esters. The non-polar ester can pass through the cell membrane and is hydrolysed by intracellular esterases to produce free fluorescein, which being polar, cannot readily pass the cell membrane and accumulates in the living cell (Rotman & Papermaster, 1966; Rotman, 1973). With this method intracellular turnover of such fluorogenic substrates can be studied (Sernetz, 1973).

Microfluorometry in microvolumes

There has been an increased use of microfluorometry in immunochemistry and enzyme chemistry for measuring FRPs in extremely small volumes of fluid contained in microdroplets, microcapillaries or microspheres of an artificial matrix. In this case the microfluorometer is only used as an instrument which can accurately measure fluorescence in 'microcuvettes' of microscopical size (diameter 1–50 μm) which cannot be handled in normal macrofluorometers.

Rotman (1961) described a method capable of measuring single molecules in droplets of microscopic size dispersed in silicon oil and visualised and measured with a microfluorometer. He measured the activity of β-D-galactosidase using as a fluorogenic substrate 6-hydroxyfluoran-β-D-galactopyranoside and could detect the activity of an individual molecule of the enzyme.

In immunofluorescence studies there is a need for methods where the amounts and concentrations of the biological reagent (antigens and antibodies) can be better defined to permit a quantitative interpretation of the immunological reactions. Special carriers of an artificial matrix with a defined and measurable microscopical dimension are used for this purpose. Van Dalen, Knapp & Ploem (1973) have introduced the use of spherical Agarose beads to which antigens are covalently bound. After such beads have reacted with fluorescein-conjugated antibodies, the resulting fluorescence is measured in a microfluorometer. In a number of applications this method appears to be more sensitive for high conjugate dilutions than immunofluorescence methods using conventional tissue sections or cells of defined antigenicity as a substrate. Deelder, Snoijink & Ploem (1975) used this method as a sensitive test for antibodies against the antigens from a human parasite (*Schistosoma mansoni*). A similar approach was followed by Bergquist, Holubar, Diaz & Beutner (1972) and Holubar, Bergquist, Lesser & Beutner (1973) using spherical particles of polymerised immunoglobulins as a substrate. Capel (1974) described a system similar to that of Van Dalen *et al.* (1973). For a review of this new method the reader should consult a review by Ploem (1975).

Galjaard *et al.* (1974) and Jongkind, Ploem, Galjaard & Reuser (1974) have described a technique for performing enzyme assays at the single cell level using microfluorometry of a microcapillary or a microdroplet containing the FRP. This technique enables enzyme determinations to be carried out when only very few cells are available, for example antenatal diagnostic tests for inborn enzyme defects.

References

Barer, R. (1964). Microscopy. In *Cytology and Cell Physiology*, ed. G. H. Bourne, pp. 91–154. Academic Press, New York & London.

Bergquist, N. R., Holubar, K., Diaz, G. A. & Beutner, E. H. (1972). The manufacture of protein microspheres by means of suspension polymerization. *International Archives of Allergy and Applied Immunology*, **43**, 791–9.

Bergquist, N. R. & Nilsson, P. (1975). Laser excitation of fluorescent copolymerized immunoglobulin beads. *Annals of the New York Academy of Sciences*, **254**, 157–62.

Björklund, A., Ehinger, B. & Falck, B. (1968). A method for differentiating dopamine in tissue sections by microspectrofluorometry. *Journal of Histochemistry and Cytochemistry*, **16**, 263–9.

Boehm, N. (1972). Fluorescence cytophotometric determination of DNA. In *Techniques of Biochemical and Biophysical Morphology*, ed. D. Glick & R. M. Rosenbaum, pp. 89–141. Wiley–Interscience, New York.

Boehm, N. & Sprenger, E. (1968). Fluorescence cytophotometry: a valuable method for the quantitative determination of nuclear Feulgen–DNA. *Histochemie*, **16**, 100–18.

Bosshard, U. (1964). Fluoreszenzmikroskopische Messung des DNS-Gehaltes von Zellkernen. *Zeitschrift für Wissenschaftliche Mikroskopie*, **65**, 391–408.

Capel, P. J. A. (1974). A quantitative immunofluorescence method based on the covalent coupling of protein to sepharose beads. *Journal of Immunological Methods*, **5**, 165–78.

Chance, B. (1962). Kinetics of enzyme reactions within single cells. *Annals of the New York Academy of Sciences*, **97**, 431–48.

Cherry, W. B. (1970). Fluorescence emission with special reference to standardization in immunofluorescence. In *Standardization in Immunofluorescence*, ed. E. J. Holborrow, pp. 127–36. Blackwell Scientific Publications, Oxford.

Coons, A. H., Creech, H. J. & Jones, R. N. (1941). Immunological properties of an antibody containing a fluorescent group. *Proceedings of the Society for Experimental Biology and Medicine*, **47**, 200–2.

Cornelisse, C. J. & Ploem, J. S. (1976). A new type of two-color fluorescence staining for cytology specimens. *Journal of Histochemistry and Cytochemistry*, **24**, 72–81.

Van Dalen, J. P. R., Knapp, W. & Ploem, J. S. (1973). Microfluorometry on antigen–antibody interaction in immunofluorescence using antigens covalently bound to agarose beads. *Journal of Immunological Methods*, **2**, 383–92.

Deelder, A. M. & Ploem, J. S. (1974). An immunofluorescence reaction for *Schistosoma mansoni* using the defined antigen substrate spheres (DASS) system. *Journal of Immunological Methods*, **4**, 239–51.

Deelder, A. M., Snoijink, J. J. & Ploem, J. S. (1975). Experimental optimization of the DASS system for immuno-diagnosis of some helminth infections. *Annals of the New York Academy of Sciences*, **254**, 119–36.

Dittrich, W. & Göhde, W. (1969). Impulsfluorometrie bei Einzelzellen in Suspensionen. *Zeitschrift für Naturforschung*, **24**, 360–1.

Dowson, J. H. & Laszlo, I. (1971). Quantitative histochemical studies of formaldehyde induced parenchymal fluorescence following L-Dopa administration. *Journal of Neurochemistry*, **18**, 2501–8.

Eränkö, O. (1964). Histochemical demonstration of catecholamine by fluorescence induced by formaldehyde vapour. *Journal of Histochemistry and Cytochemistry*, **12**, 487–9.

Falck, B., Hillarp, N. A., Thieme, G. & Torp, A. (1962). Fluorescence of catecholamines and related compounds condensed with formaldehyde. *Journal of Histochemistry and Cytochemistry*. **10**, 348–54.

Fulwyler, M. J. (1974). Status quo in flow-through cytometry. *Journal of Histochemistry and Cytochemistry*, **22**, 605–6.

Galjaard, H., Hoogeveen, A., Keijzer, W., De Wit-Verbeek, E. & Vlek-van Noot, C. (1974). The use of quantitative cytochemical analysis in

rapid prenatal detection and somatic cell genetic studies of metabolic diseases. *Histochemical Journal*, **6**, 491–509.
Gill, J. E. & Jotz, M. M. (1974). Desoxyribonucleic acid cytochemistry for automated cytology. *Journal of Histochemistry and Cytochemistry*, **22**, 470–7.
Goldman, M. (1960). Antigenic analysis of *Entamoeba histolytica* by means of fluorescent antibody. I. Instrumentation for microfluorimetry of stained amebae. *Experimental Parasitology*, **9**, 25–36.
Goldman, M. (1967). An improved microfluorometer for measuring brightness of fluorescent antibody reactions. *Journal of Histochemistry and Cytochemistry*, **15**, 38–45.
Goldman, M. (1968). *Fluorescent Antibody Methods*. Academic Press, New York & London.
Goldman, M. (1971). Microfluorimetry in the antigenic analysis of closely related microorganisms. *Annals of the New York Academy of Sciences*, **177**, 439–45.
Haag, D. (1973). Untersuchungen zur Proportionalität zwischen DNS-Gehalt und Fluoreszenzintensität fluorochromierten Einzelzellen. *Histochemie*, **36**, 283–91.
Haaijman, J. J. & Van Dalen, J. P. R. (1974). Quantification in immunofluorescence microscopy. A new standard for fluorescein and rhodamine emission measurement. *Journal of Immunological Methods*, **5**, 359–74.
Holubar, K., Bergquist, N. R., Lesser, G. R. & Beutner, E. H. (1973). Comparative microfluorometric studies of activity and size of suspension polymerized anti-IgG antibodies at different pH levels. *International Archives of Allergy and Applied Immunology*, **44**, 489–99.
Irvine, W. J., Chan, M. M. W. & Williamson, W. G. (1969). The application of microphotometry to the indirect immunofluorescent antibody technique using the human gastric parietal cell antigen–antibody system. *Clinical and Experimental Immunology*, **4**, 607–17.
Jongkind, J. F., Ploem, J. S., Galjaard, H. & Reuser, A. J. J. (1974). Enzyme assays at the single cell level using a new type of microfluorimeter. *Histochemistry*, **40**, 221–30.
Jongsma, A. P. M., Hijmans, W. & Ploem, J. S. (1971). Quantitative immunofluorescence. Standardization and calibration in microfluorometry. *Histochemie*, **25**, 329–43.
Jonsson, G. (1971). Quantitation of fluorescence of biogenic monoamines. *Progress in Histochemistry and Cytochemistry*, **2**, 299–333.
Kaufman, G. I., Nester, J. F. & Wasserman, D. E. (1971). An experimental study of lasers as excitation sources for automated fluorescent antibody instrumentation. *Journal of Histochemistry and Cytochemistry*, **19**, 469–76.
Killander, D., Hellström, U., Johansson, N., Klein, E., Levin, A. & Perlmann, P. (1973). Immunofluorometric determinations of surface immunoglobulins of individual normal and malignant human lymphoid cells. In *Fluorescence Techniques in Cell Biology*, ed. A. A. Thaer & M. Sernetz, pp. 163–81. Springer-Verlag, Berlin.
Kohen, E., Kohen, C., Thorell, B. & Wagener, G. (1973). Quantitative aspects of rapid microfluorimetry for the study of enzyme reactions and transport mechanisms in single living cells. In *Fluorescence Techniques in Cell Biology*, ed. A. A. Thaer & M. Sernetz, pp. 207–18. Springer Verlag, Berlin.

Kohen, E., Thorell, B., Kohen, C. & Salmon, J. M. (1974). Studies on metabolic events in localized compartments of the living cell by rapid microspectrofluorometry. In *Advances in Biological and Medical Physics*, ed. J. H. Lawrence & J. W. Goffman, vol. 15, pp. 271–94. Academic Press, New York & London.

Kolk, A. H. K., Samuel, T. & Ruemke, P. (1974). Auto-antigens of human spermatozoa. I. Solubilization of a new auto-antigen detected on swollen spermheads. *Clinical and Experimental Immunology*, **16**, 63–76.

Kraft, W. (1970). Ein neues FITC-Erregerfilter für die Routinefluoreszenz. *Leitz-Mitteilungen Wissenschaft und Technologie*, **5** (2), 41–4.

Kraft, W. (1975). The technology of new fluorescence illumination systems. *Mikroskopie*, **31**, 129–46.

Levin, A., Killander, D., Klein, E., Nordenskjöld, B. & Inoue, M. (1971). Applications of microspectrofluorimetry in quantitation of immunofluorescence on single cells. *Annals of the New York Academy of Sciences* **177**, 481–9.

Lichtensteiger, W. (1970). Katecholaminhaltige Neurone in der Neuroendokrinen Steuerung. Prinzip und Anwendung der Mikrofluorimetrie. *Progress in Histochemistry and Cytochemistry*, **1**, 185–276.

Loken, M. R. & Herzenberg, L. A. (1975). Analysis of cell populations with a fluorescence-activated cell sorter. *Annals of the New York Academy of Sciences*, **254**, 163–71.

Mansberg, H. P. & Kusnetz, J. (1966). Quantitative fluorescence microscopy: fluorescent antibody automatic scanning techniques. *Journal of Histochemistry and Cytochemistry*, **14**, 260–73.

Mayall, B. H. & Mendelsohn, M. L. (1970). Errors in absorption cytophotometry: some theoretical and practical considerations. In *Introduction to Quantitative Cytochemistry*, vol. 2, ed. G. L. Wied & G. F. Bahr, pp. 171–97. Academic Press, New York & London.

Nairn, R. C. (1968). Standardization in immunofluorescence. *Clinical and Experimental Immunology*, **3**, 465–76.

Nairn, R. C. (1969). *Fluorescent Protein Tracing*. E. & S. Livingstone, Edinburgh.

Nairn, R. C., Herzog, F., Ward, H. & de Boer, W. G. R. M. (1969). Microphotometry in immunofluorescence. *Clinical and Experimental Immunology*, **4**, 697–705.

Nitsch, B., Murken, J. D. & Brück, H. J. (1970). Determining Feulgen-DNA of individual chromosomes by fluorescence cytophotometry with incident light. *Histochemie*, **23**, 254–65.

Ornstein, L. (1952). The distributional error in microspectrophotometry. *Laboratory Investigations*, **1**, 250–65.

Pearse, A. G. E. (1972). *Histochemistry. Theoretical and Applied.* Churchill Livingstone, Edinburgh & London.

Le Pecq, J. P. (1973). Ethidium bromide: a fluorescent probe of nucleic acid structure and its potential for in vivo studies. In *Fluorescence Techniques in Cell Biology*, ed. A. A. Thaer & M. Sernetz, pp. 301–9. Springer-Verlag, Berlin.

van der Ploeg, M. & Ploem, J. S. (1972). Filter combinations and light sources for fluorescence microscopy of quinacrine mustard or quinacrine stained chromosomes. *Histochemie*, **33**, 61–70.

Ploem, J. S. (1967). The use of a vertical illuminator with interchangeable dichroic mirrors for fluorescence microscopy with incident light. *Zeitschrift für wissenschaftliche Mikroscopie*, **68**, 129–42.

Ploem, J. S. (1970*a*). Quantitative immunofluorescence. In *Standardization in*

Immunofluorescence, ed. E. J. Holborow, pp. 63–73. Blackwell Scientific Publications, Oxford.

Ploem, J. S. (1970*b*). Standards for fluorescence microscopy. In: *Standardization in Immunofluorescence*, ed. E. J. Holborow, pp. 137–53. Blackwell Scientific Publications, Oxford.

Ploem, J. S. (1971). A study of filters and light sources in immunofluorescence microscopy. *Annals of the New York Academy of Sciences*, **177**, 414–29.

Ploem, J. S. (1973). Immunofluorescence microscopy. In *Immunopathology of the Skin: Labeled Antibody Studies*, ed. E. H. Beutner, pp. 248–70. Dowden, Hutchinson & Ross Inc., Stroudsburg, Penn.

Ploem, J. S. (1975). General introduction. *Annals of the New York Academy of Sciences*, **254**, 4–20.

Ploem, J. S., de Sterke, J. A., Bonnet, J. & Wasmund, H. (1974). A microspectrofluorometer with epi-illumination operated under computer control. *Journal of Histochemistry and Cytochemistry*, **22**, 668–77.

Prenna, G., Leiva, S. & Mazzini, G. (1974*a*). Quantitation of DNA by cytofluorometry of the conventional Feulgen reaction. *Histochemical Journal*, **6**, 467–89.

Prenna, G., Mazzini, G. & Cova, S. (1974*b*). Methodological and instrumentational aspects of cytofluorometry. *Histochemical Journal*, **6**, 259–78.

Richards, O. W. & Waters, P. (1967). A new interference exciter filter for fluorescence microscopy of fluorescein-tagged substances. *Stain Technology*, **42**, 320–1.

Rigler, R. (1966). Microfluorometric characterization of intracellular nucleic acids and nucleoproteins by Acridine Orange. *Acta Physiologica Scandinavica* (*Supplement*), **267**, 1–122.

Rost, F. W. D. (1972). Fluorescence microscopy. In *Histochemistry, Theoretical and Applied*, ed. A. G. E. Pearse, vol. 2, pp. 1171–1206. Churchill Livingstone, Edinburgh & London.

Rost, F. W. D. & Pearse, A. G. E. (1971). An improved microspectrofluorimeter with automatic digital data logging: construction and operation. *Journal of Microscopy*, **94**, 93–105.

Rotman, B. (1961). Measurement of activity of single molecules of β-D-galactosidase. *Proceedings of the National Academy of Sciences, USA*, **47**, 1981–91.

Rotman, B. (1973). Changes in the membrane permeability of human leukocytes measured by fluorochromasia in a rapid flow fluorometer. In *Fluorescence Techniques in Cell Biology*, ed. A. A. Thaer & M. Sernetz, pp. 255-8. Springer-Verlag, Berlin.

Rotman, B. & Papermaster, B. W. (1966). Membrane properties of living mammalian cells as studied by enzymatic hydrolysis of fluorogenic esters. *Proceedings of the National Academy of Sciences, USA*, **55**, 134–41.

Ruch, F. (1964). Fluoreszenzphotometrie. *Acta Histochemica* (*Supplement*), **6**, 117–21.

Ruch, F. (1966). Determination of DNA content by microfluorometry. In *Introduction to Quantitative Cytochemistry*, ed. G. L. Wied, pp. 281–94. Academic Press, New York & London.

Ruch, F. (1970). Principles and some applications of cytofluorometry. In *Introduction to Quantitative Cytochemistry*, vol. 2, ed. G. L. Wied & G. F. Bahr, pp. 431–50. Academic Press, New York & London.

Ruch, F. & Bosshard, U. (1963). Photometrische Bestimmung von Stoffmenge im Fluoreszenzmikroskop. *Zeitschrift für Mikroscopie* **65**, 335–41.

Ruch, F. & Leeman, U. (1973). Cytofluorometry. In *Molecular Biology*

Biochemistry and Biophysics, vol. 14, *Micromethods in Molecular Biology*, ed. V. Neuhof, pp. 329–46. Springer-Verlag, Berlin.

Rygaard, J. & Olson, W. (1969). Interference filters for improved immunofluorescence microscopy. *Acta pathologica et microbiologica Scandinavica*, **76**, 146–8.

Sernetz, M. (1973). Microfluorometric investigations on the intracellular turnover of fluorogenic substrates. In *Fluorescence Techniques in Cell Biology*, ed. A. A. Thaer & M. Sernetz, pp. 243–54. Springer-Verlag, Berlin.

Sernetz, M. & Thaer, A. A. (1970). A capillary fluorescence standard for microfluorometry. *Journal of Microscopy*, **91**, 43–52.

Steinkamp, J. A. & Crissman, H. A. (1974). Automated analysis of deoxyribonucleic acid, protein and nuclear to cytoplasmic relationships in tumor cells and gynecologic specimens. *Journal of Histochemistry and Cytochemistry*, **22**, 616–21.

Taylor, C. E. D., Heimer, G. V. & Lidwell, O. M. (1971). Use of a fibre optic probe for quantitative immunofluorescence studies *The Lancet*, ii, 785–6.

Taylor, C. E. D. & Heimer, G. V. (1975). Quantitative immunofluorescence studies. *Annals of the New York Academy of Sciences*, **254**, 151–6.

Tilders, F. J. H., Ploem, J. S. & Smelik, P. G. (1974). Quantitative microfluorometric studies on formaldehyde-induced fluorescence of 5-hydroxytryptamine in the pineal gland of the rat. *Journal of Histochemistry and Cytochemistry*, **22**, 967–75.

Tomlinson, A. H. (1967). Filters for use with an iodine–quartz lamp to excite immunofluorescence. *Immunology* **13**, 323–4.

Tomlinson, A. H. (1970). An interference filter for use with the iodine–quartz lamp to excite fluorescence. In *Standardization in Immunofluorescence*, ed. E. J. Holborow, pp. 117–21. Blackwell Scientific Publications, Oxford.

Tomlinson, A. H. (1971). Tungsten halogen lamps and interference filters for immunofluorescence microscopy. *Proceedings of the Royal Microscopical Society*, **7**, 27–37.

Wachsmuth, E. D. & Woodhams, R. L. (1974). A quantitative approach for immunofluorescence in microscopy: the use of antibody multilayers on nuclei. *Histochemistry*, **38**, 339–50.

Walter, F. (1970). Fluoreszenzmikroskopie in Biologie und Medizin. *Leitz-Mitteilungen Wissenschaft und Technologie*, **5** (2), 33–40.

Wasmund, H. (1976). Computergesteuerte Scanning-Mikroskop für die Analyse von Einzelzellen. *Leitz-Mitteilungen Wissenschaft und Technologie*, **6** (6), 217–23.

West, S. S. (1969). Fluorescence microspectrophotometry of supravitally stained cells. In *Physical Techniques in Biological Research*, vol. III, part C, *Cells and Tissues*, ed. A. W. Pollister, pp. 253–317. Academic Press, New York & London.

Wick, G., Schauenstein, K., Herzog, F. & Steinbatz, A. (1975). Investigations of the recovery phenomenon after laser excitation in immunofluorescence. *Annals of the New York Academy of Sciences*, **254**, 172–4.

Young, M. R. (1961). Principles and technique of fluorescence microscopy. *Quarterly Journal of microscopical Science*, **102**, 419–49.

S. BRADBURY

Quantitative image analysis

Many features of the microscope image, especially those related to shape and pattern, can be appreciated easily by the eye and brain of a trained observer. This has been the way in which the microscope has been used in biology for over three hundred years. In the last decade or so, however, there have been tremendous advances in the use of the microscope for the acquisition of numerical data from the image. Collection of such data has been possible for many years using the point-counting methods which form the basis of manual stereological techniques; these are slow and laborious but do provide objective data.

Developments in television scanning techniques and in electronics have now made the image analyser a practicable proposition and systems are currently available from several manufacturers. Such TV-based image analysers give speed and accuracy to measurements in routine situations such as the assessment of response of goblet cells in the tracheal mucosa to an irritant gas in the inspired air. In experimental situations the image-analysing computer provides objective data on many cell parameters which may then be used as discriminators of experimentally induced changes. The numerical data thus acquired allow the use of powerful statistical methods in morphological situations where they would not otherwise be applicable. Large numbers of specimens can be analysed, measurements (e.g. integrated optical densities) not normally obtainable with point-counting methods can be used, and observer error due to boredom and fatigue much reduced.

The Quantimet 720 system

The Quantimet 720 (QTM 720) shown in Fig. 1 is a modular system capable of expansion at any time as all new modules are compatible with previous units; each purchaser can thus select those modules most appropriate to the problems he has, but retain the option of changing or updating his system at any time. With any given combination of modules several possibilities exist

Fig. 1. A large Quantimet 720 system. From left to right can be seen the 9830 calculator, microscope deck with lamp power supplies and the scanner camera connected to the microscope by a prism, and the display deck and processor modules.

in the manner in which they are interconnected (the so-called 'high level programming'); this greatly increases the power and flexibility of any given system. A good introductory review of the QTM 720 system is given by Fisher (1971).

Any system must contain input peripherals, central processor modules and output peripherals to handle the data generated by the processor. In the majority of modern systems a large programmable calculator or small extra computer is linked to an output data interface in order to process further the numerical data generated by the QTM 720. A summary of the modules currently available in the system is given in Table 1, and some possible ways in which they may be interconnected are indicated in the block diagram (Fig. 2).

Input peripherals

The image needed by the QTM scanner may be provided directly from an optical or electron microscope (either transmission or scanning), whilst for viewing objects of up to 8 cm × 10 cm in size an epidiascope is available; this may be fitted with one of a selection of lenses of differing focal length

Table 1. *A summary of some of the QTM 720 modules and their principal functions*

Module		Main function
1	Microscope stage control	Stepping of microscope stage under programmer control
2	Orientation control	Rotation of image under programmer control
3	Focus control	Used with 1 above; selects best focal plane automatically
4	System control	Provides scanner control, video processing and generation of timing pulses, data accumulation and character display, shading correction
5	Display unit	Image display
6	Standard detector	Enables operator selection of required features in terms of grey scale between black and white
7	1D autodetector	Automatically detects feature edges at 50 % amplitude level on a one-dimensional basis
8	2D autodetector	Automatically detects feature edges at 50 % amplitude level on a two-dimensional basis
9	Densitometer	Provides logarithmic equivalent of scanner signal to give integrated optical density of field or feature
10	SEM interface and detector	Allows operator to select wanted features on basis of electron image contrast
11	Amender	Enables operator to make controlled variations to output from detector module in the scan direction
12	2D amender	Enables operator to dilate or erode image output from detector in both horizontal and vertical directions by expanding or contracting the perimeter of the detected image
13	MS3 computer	Processes detected or amended video to give the four basic measurements: provides chord sizing; allows measurements on modified video; controls pattern recognition modules
14	Function computer	Operates in conjunction with MS3 to provide feature-specific measurements
15	Form separator	Computes form function for each feature in turn from data provided by function computers
16	Classifier/collector	(i) Accepts data from function computer and compares with pre-set limits; classifies features in terms of any of the F/C parameters (ii) Collects gating signals from MS3

Table 1 (*cont.*)

Module	Main function
	computer form separator and classifier to identify those features which satisfy discriminating criteria
17 Size distributor	Allows seven extra sizing channels to be added to system
18 Variable frame and scale	Produces a variable size frame to serve as the measuring area instead of the standard frame generated by the system control; provides a display of a calibration scale
19 Programmer	Steps machine through a predetermined sequence of switchings and measurements
20 Control interface	Control module for use with a large calculator to provide software programme control of QTM sequencing and measurements (replaces 19 above)
21 Triple supervisor	Exercises surveillance over data whose selection and entry are controlled by programmer; indicates when any level of data exceeds, is less than or is equal to present values
22 Multiple accumulator	Contains eight data accumulating stores, the contents of which are displayed in turn on the screen
23 Light pen	Used for operator interaction with the displayed image to obtain feature-specific information on a single feature
24 Image editor	Interactive memory light pen which allows more elaborate operator interaction with image than 23 above
25 Calculator field data interface	Accepts whole field data from QTM processor and transfers it to selected locations in the memory of a programmable desk calculator
26 Calculator field/ feature interface (CFFI)	Transfers QTM field data from MS3 or classifier/collector module to calculator; transfers feature data for each feature in turn to calculator; co-operates with function computer in the measurement of 45° and 135° Feret diameters and co-ordinates of individual features
27 Field image feature interface (FIFI)	Allows QTM data to be transferred on a feature, whole field or full image basis to a PDP11 computer for further processing; allows interactive QTM/computer display facilities on QTM screen
28 Teletype drive	Accepts data from QTM processor which is then transferred to teletype for printing and tape punching; in modified form can accept feature data from CFFI for printing and punching

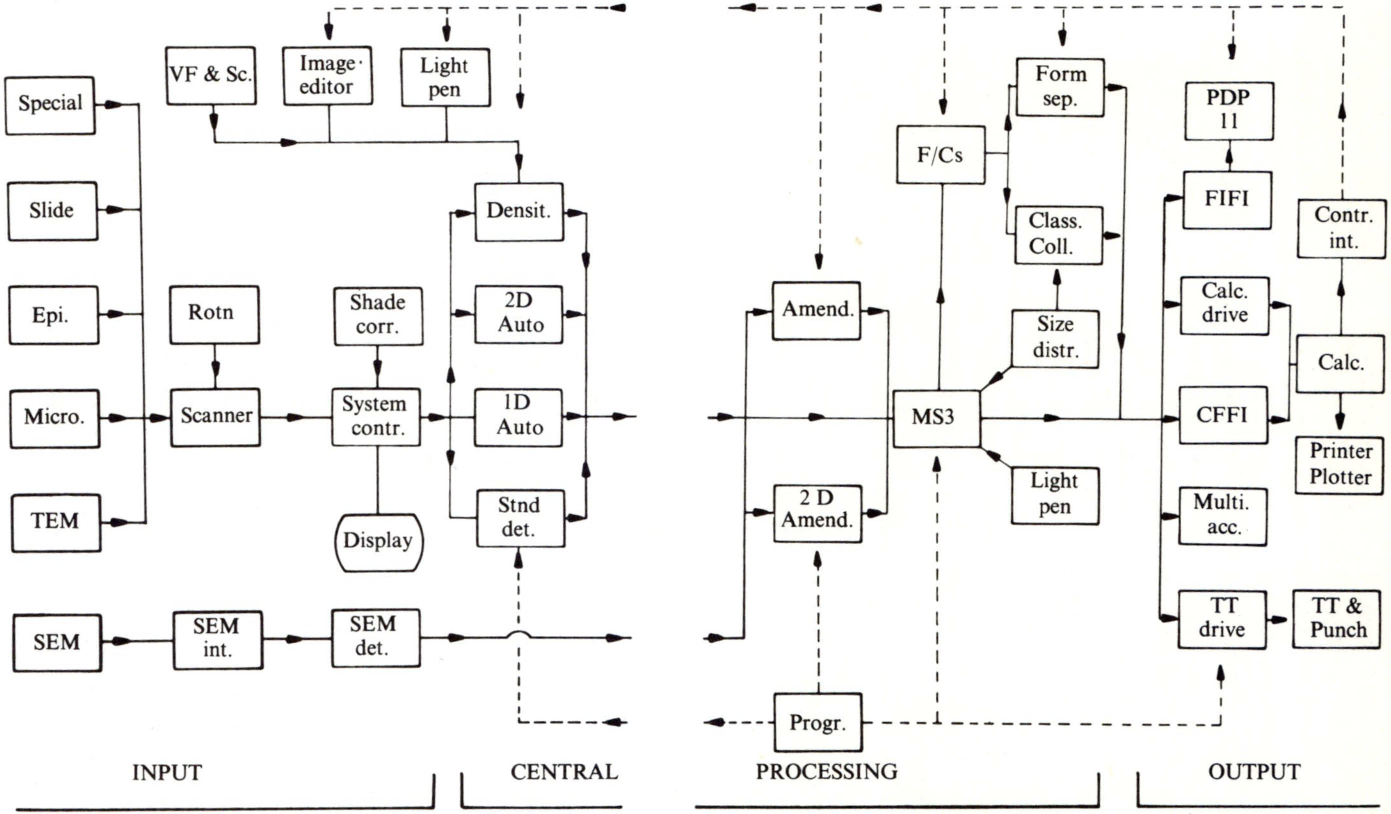

Fig. 2. A block diagram of the modules which form the QTM720 system. Control pathways from the programmer or control interface are shown as dashed lines. Note: not all of these modules would be available in any given system. For example, the 1D autodetector might be the only detection module present in the system; alternatively, the standard detector might be in use together with a densitometer.

Key to abbreviations. Amend., amender; 2D Amend., 2D amender; Calc., programmable caculator; Calc. drive, calculator drive interface; CFFI, calculator field feature interface; contr. int., control interface; Class. Coll., classifer/collector; 2D Auto, 2D autodetector; 1D Auto, 1D autodetector; Densit., densitometer; Epi., epidiascope input; F/Cs, function computers; FIFI, field image feature interface; Form sep., form separator; micro., microscope input; Multi. acc., multiple data accumulator; Progr, programmer; Rotn, rotation of image control; SEM, scanning electron microscope input; SEM int., scanning EM interface; SEM det., scanning EM detector; Shade corr., shading corrector; Size distr., size distributor; Slide, slide scanning input; Special, special imaging systems; Stnd det., standard detector; System contr., system control; TT, teletype; TT drive, teletype drive; TEM, transmission electron microscope input; VF & Sc., variable frame and scale.

so as to obtain the most appropriate magnification for any given application. Equipment for viewing 35-mm slides, 16-mm and 35-mm cine film is also available. Because of the complexity of the images of many biological objects, the epidiascope has hitherto been regarded as essential for any QTM system used in biological work.

The procedure has been to prepare a drawing or photograph of the object and then to enhance or suppress detail until a measurable image is obtained. Such a way of working (known as an 'interactive approach') was slow, and open to operator bias. Recently the introduction of the image editor module has rendered this type of approach much less necessary. One disadvantage of the epidiascope attachment is the limitation on the maximum size of picture which may be examined. To overcome this I have constructed a holder for large pictures (up to 12″ × 10″) using a wall-mounted scanner bracket and picture holder fitted with fluorescent tubes to provide an even illumination on the picture holder.

The scanner contains either a vidicon or plumbicon tube. The vidicon is more robust, with a longer life, and serves as the general purpose camera. If, however, a plumbicon is used there is the advantage that it is possible to integrate the light over several field scans. This means that phase contrast, Nomarski interference contrast and incident-light dark-ground may be used. The latter technique is especially valuable for the quantification of autoradiographs. A plumbicon camera also allows the use of fluorescence microscopy with the low light intensities inherent in this method. If much densitometry is contemplated, a plumbicon tube is advisable due to its greater sensitivity and the ease with which the system may be calibrated in units of absolute optical density. The scanner is constructed to rigorous standards and uses a scan standard of 720 lines scanned sequentially 10.5 times per second. This gives an equal resolution both along and at right-angles to the scan lines and is the best possible compromise between noise, resolution and speed. In addition considerable care is taken to achieve uniformity over the whole field which is broken up for measurement into over 605 440 separate 'picture points'. The final compensation for non-uniformity in the field ('shading') is achieved in the system control module by an electronic 'shading corrector' which reduces such errors to less than 1 %. The system control module forms an essential part of even the smallest QTM 720 system; it has three functions, namely to control the scanner and video processing, to generate the timing and synchronisation pulses for the whole instrument and to accumulate data and generate characters for display on the monitor screen. Video signals from the system control and other modules are displayed on a television monitor tube so that the image may be observed and regions which are being operated on may be brightened up by various types of characteristic display.

Processor modules

From the diagram of the instrument (Fig. 2) it may be seen that this group of modules comprises the detectors, the central MS3 computer and the various modules (function computers, form separator and classifier/collector) which form the so-called 'pattern recognition' system. Some type of detector is essential in any QTM system to assess separately every picture point in terms of its grey value. The detector allows the operator to select features in the field of view for measurement whilst at the same time ignoring unwanted features of a different optical density. Three continuously variable threshold setting controls are available, each one of which can be set anywhere between black and white, so that four different grey levels may be chosen for measurement at any setting of the three controls. The measurement always occurs within a clearly defined area – the 'live frame', which is shown on the display tube together with a bright-up of the detected features. The standard detector has now been superseded by the 1D and 2D autodetectors which are designed to overcome certain technical setting errors (offset threshold setting error and halo error) inherent in the standard detector; a full account of these errors is given in the paper by Wadlow, Hopkins, Gardner & Fisher (1972) together with an account of the principles underlying autodetection. In effect, a memory unit is used so that the detector 'sees' not only the current picture point but also a number of surrounding picture points, either in a line (1D) or as a matrix (2D). This line or matrix is dynamic and moves across the field with the scan spot; in consequence, the detector can judge the significance of the grey level of each picture point in relation to its neighbours and so tell whether it is one of a large area of the same grey value or is in a rapidly changing region between two phases. In the latter case the detector module can observe the value of both grey phases away from the boundary region and so decide automatically where the 50 % level of correct detection occurs.

It is also possible to install a densitometer module into the system. This converts the scanner video signal to its logarithmic equivalent and then digitises this into 6-bit, 64 levels which can be calibrated in absolute optical density units. This module thus gives as its primary output the integrated optical density of a feature or field, i.e. the sum of the density contribution of each individual detected picture point. As the detected area may also be obtained, the mean density is readily derived if required. The densitometer module has an enormous potential, especially in the field of nuclear DNA measurement and in the quantitation of enzyme histochemical reactions (Bond, 1973).

The heart of the central processor is the standard MS3 computer which accepts the detected and amended or edited video signal; this is then processed to obtain the four basic measurements (see below). If the MS3 is in the so-called 'pattern recognition' mode it also controls further modules known as function computers (or function analysers), a form separator and a classifier/collector. These allow feature (as opposed to full field) measurements and so a classification and analysis using a number of separate measurements of different parameters on each feature may take place. The function computers allow the computation of area, perimeter, horizontal and vertical intercepts and horizontal and vertical Feret diameters for each feature at full speed (10.5 frames s^{-1}) no matter how many features are present in the field.

The form separator is used to compute, for each feature separately, the form function $A \times B/C^n$, where A, B and C are any three parameters measured by the function computers and n has the value of 1, 2 or 3 according to the position of a pre-set switch. This function is a non-dimensional, orientation-independent form factor; examples of some of these are given in Table 2 and considered in more detail later. As pre-set limit switches are included on the module it is possible to ensure that only the data from those features which satisfy certain form factors are passed on for further evaluation. The form separator module thus allows an elementary classification of features on a shape basis. It should be noted, however, that with the development of more sophisticated calculators used on-line to the QTM 720, the computation of shape discriminators for individual features is probably done more easily in the calculator.

The classifier/collector is really two modules in one, each with separate functions. As its name implies, the classifier accepts the data from a function computer and compares it with pre-set or pre-programmed limits; this allows features to be further classified in terms of any of the functions measurable by the computer. The collector gathers together signals from the function computers, classifier and form separator and identifies those features in the field being examined which satisfy all the discriminating criteria; feature data from these only is released to the output modules.

Output peripherals

In a simple QTM system, the data is presented in a numerical form at the top of the display screen from which they may be noted down for further manual processing by the operator. With the increase in complexity of the QTM system and with the possibility of using feature data rather than summed field data, the need for more sophisticated data handling modules becomes apparent. These may take the form of a teletype drive interface or of complex calculator interfaces. The teletype drive allows the data from the

Table 2. *Some arbitrary form separation factors and derived parameters which are of use as shape discriminators*

Form factor or derived parameter	Use
A/P^2 or $\sqrt{(A/P)}$	Circularity function
Max F/Min F	Elongation factor
Mean F	Convex perimeter of 'envelope' of object
VF/HF	Orientation of ordered objects
HF/VF	Degree of re-entrance of ordered features
2A/P	Mean width
A/P	Mean chord (averaged over all orientations)
V/A	Mean density
A1/A2	Two phase ratio
	Specific cases for fibres
P/2	Length for straight fibres
$P+(P^2-16A)^{\frac{1}{2}}/4$	L1 length approximation for fibres (sensitive to perimeter)
F Max	L2 length approximation for straight fibres
$P-(P^2-16A)^{\frac{1}{2}}/4$	Width approximation for fibres
A/F Max	Width approximation for straight fibres
L/W	Aspect ratio of a fibre
$L1/L2 = \dfrac{P+(P^2-16A)^{\frac{1}{2}}/4}{F\ Max}$	Re-entrancy factor, used to eliminate crossed fibres, particle clusters etc.

Key:

A, area; P, perimeter; V, integrated optical density; F, Feret diameter (maximum, minimum or mean); VF, vertical Feret diameter; HF, horizontal Feret diameter; L, calculated length; W, calculated width.

QTM to be printed and, at the same time, reproduced in the form of a punched paper tape. This may be easily stored and the data reprocessed at any future date by running the tape through the reader attached to a calculator such as the Hewlett Packard 9810 or 9830. This method of operation has the great advantage that the data from a single measuring run are preserved; if a large number of parameters from each individual feature are taken onto the tape then only those items relevant to each particular calculation need be used and the remainder of the data is available for subsequent processing at any future date without the need for remeasuring the object on the Quantimet. This is especially useful in the early stages of an investigation when it may not be clear which form separators are going to be meaningful and again when in the later stages of a study some derived parameter is required. Taking the raw data onto paper tape does, however, suffer from the disadvantage that the standard tape punch on the teletype is rather slow in operation: if many features are to be measured and if, say, seven individual measurements are to be made on each feature, then the recording of the data becomes rather tedious. Interfacing a high-speed tape punch alone to the QTM 720 is obviously the solution to this problem.

The need for more and more powerful calculation facilities has now become accepted. A simple calculator drive interface is available for the Quantimet but much more useful is the calculator field-feature interface (CFFI) which not only contains a calculator interface presenting the numerical data in a suitable form to the calculator of choice, but also allows the serial acquisition of a chosen list of feature data from each of those features in a field, in turn, which satisfy any acceptance criteria pre-set on the computer or pattern recognition modules. The CFFI module is also capable of modifying the function computer logic to provide the Feret diameters at 45° and 135° to the normal Feret diameters which are available on any standard Quantimet. For biologists this is a very useful module used in conjunction with a large programmable calculator, when it becomes easy to sort and process feature data and present size distribution histograms and density distribution histograms which have been normalised and for which all the basic statistics have been calculated. Purchasers of QTM systems always find after one or two years that as their expertise and ambition in image analysis grow they have insufficient calculating power; a point to be remembered in the initial stages of planning an installation. If only limited funds are available then it is probably preferable to spend them on the data processing chain of modules rather than on central processor modules which may have only a limited use in biological applications where the images to be handled are complex and as such do not lend themselves to the simple shape classifications which are provided by hardware.

The most powerful data processing system of all is the linkage of a PDP11 computer to the Quantimet 720 by means of the field-image-feature interface (FIFI). This interface can transmit the data in the image to the PDP computer on a field-by-field basis, on a feature-by-feature basis, or full image data may be transferred to the computer for a more detailed analysis of chosen areas of the image. This interface is valuable where large numbers of features must be analysed (because of its speed) but it also has the great advantage that it gives the possibility of an interactive system. Numerical data are normally transferred from the QTM to the PDP11, but numerical data or alphabetical labels may now be transferred in the reverse direction to be displayed on the QTM screen against the feature to which they relate; again, statistical information or continuously updated histograms may be displayed on the QTM screen to give the operator an opportunity to judge whether sufficient features or fields have been analysed. An interesting example of the use of this module is given by Fraser (1975) in a report of a study of the development of a programme for the automatic identification and analysis of asbestos fibres.

Accessory modules

Many accessory modules exist which, although not a part of the main sequence from the input to the output, nevertheless greatly aid the image analysis process. Some, such as the variable frame and scale, the light pen and the image editor, are concerned with simplifying the delineation and acceptance of relevant features. In biological work it is not easy to obtain an analysable image using grey levels alone; some improvement may often be effected by developing special forms of material preparation, e.g. special stains of high specificity.

It is sometimes possible to use the variable frame and scale module to improve selectivity of the system by restricting the live frame (the area in which actual measurement takes place) to the area immediately surrounding the feature of interest. This works with whole field measurements but is very slow if individual features in a field are being measured. Alternatively, the use of the light pen allows the operator to interact directly with the screen image and select a specific feature from which data will be computed and passed to the output modules.

One of the more recent modules which has greatly simplified the use of the Quantimet in biological fields is the 'image editor', which uses a tracing light pen to allow the operator to interest with the displayed image on the monitor. The module thus allows the acceptance or rejection of detected features or the separation of touching features when it is set in the 'edit' mode; in the alternate 'draw' mode the operator can outline features which cannot for any reason be detected. Any number of features in a field can be operated upon with the editor and the final image may then be measured in either the field or the individual feature mode. The operation of this module is illustrated in Fig. 3. A further facility provided by the image editor is the ability to store a fully edited image in the memory of the module and subsequently to interact it with a second, different image to obtain a measure of the points of difference.

Images may also be treated with the 2D amender module to modify their detail by the processes known as 'dilation' and 'erosion'. Dilation enables the operator to agglomerate close features by expanding the image detail. This is done by moving the perimeter of the detected image outwards by a definite amount. In theory this could best be realised geometrically by adding a circular amendment element around the outside of the boundary of the detected feature. With the QTM, however, as the picture is made up of a

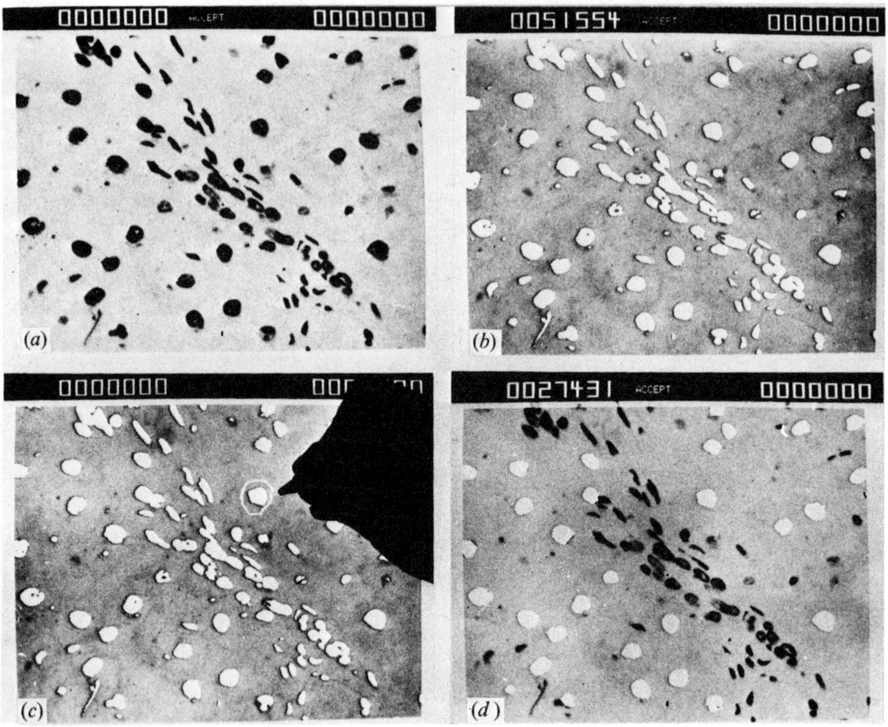

Fig. 3. A sequence of photographs of the QTM display screen to illustrate the operation of the editing of a field. (*a*) The displayed image of a field of liver cell and connective tissue nuclei. (*b*) The same field with all the nuclei detected and the instrument set to measure their total area, which is indicated as 51 554 picture points. (*c*) Image editing in progress; the tracing light pen is seen drawing a ring around a nucleus which is to be measured. (*d*) The fully edited field; only hepatocytes have been accepted (as indicated by the bright display), connective tissue nuclei now being ignored. The total area measurement (27431 picture points) is now 53 % of the former total.

rectilinear array of picture points this cannot be realised and an octagonal dilation or erosion element is used as a compromise. Dilation allows groups of objects to be measured rather than their individual fragments; this allows cell nuclei to be counted as a function of their closest neighbour distance in any direction irrespective of any anisotropy in order to obtain an estimate of the closeness of their packing. The converse process ('erosion') is used to separate touching image detail or to eliminate the narrow parts of features. In a large system one would expect to operate with two 2D amenders in series so that the operations of 'opening' (erosion followed by dilation) and 'closing' (dilation followed by erosion) may be used.

It is obvious that with a large QTM system some form of automatic control is needed in order to exploit to the full the potentialities of the inherent speed of the instrument. This control may be provided either by the standard programmer module or by the new 'control interface'. The programmer allows the machine to run through a predetermined series of measurements set by means of switched diodes arranged on a board to form a matrix of sixteen rows of thirty-two columns. This means that up to thirty-two separate instructions can be programmed for each field or feature, using any of the modules in the system which are wired in to the programmer. Such a system has limitations and with the increasing use of large calculators 'on-line', it seems logical to use the programme in the calculator to generate the instructions and sequencing for the image analyser as well as serving for the data processing. This is done with the control interface. There are many advantages to using a calculator to control QTM operation. Only one programme (that for the calculator) need be written and loaded for any problem, size distributions may be done more easily as manually set limits are not involved, and the range and linearity of the limits may be changed at will. One further advantage is that the control interface can intermingle the acquisition of both field and feature data as necessary. We may need to obtain a size distribution of feature areas and (for the same field) to obtain a total area fraction; the former requires the feature mode whilst the area fraction measurement requires the field mode of operation. To do this with the standard programmer involves many redundant measurements as the field measurements (which are only required once for each field) are of necessity repeated for each individual feature in the field. Other technical advantages also exist and it seems likely that this module will be the standard method of controlling any large QTM system.

Data output from the QTM 720

Basic measurements

The basic measurements from the computer are area, intercept, perimeter and count. All these, save for perimeter, are subject to any chord-sizing provisions manually pre-set in the computer. The standard computer normally sums the measurements from each individual detected feature in the live frame and thus provides a number which is a 'field' measure. It is possible, however, as mentioned above, to use either a light pen or the CFFI module and a function computer to obtain measurements from each individual feature in the field in turn. The data are expressed by the machine in terms of 'picture points'; as the value in absolute terms of a picture point varies according to the

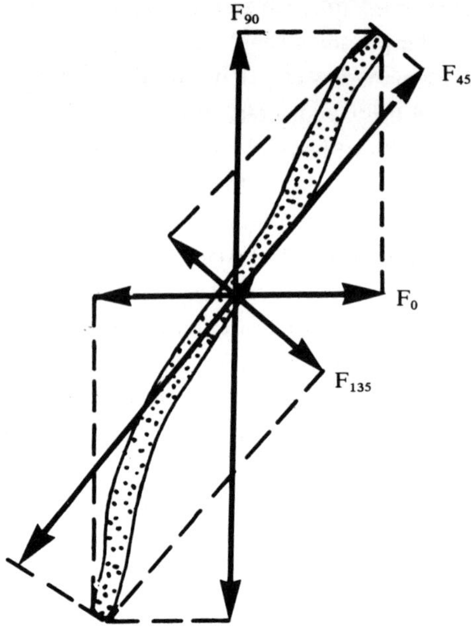

Fig. 4. A diagram of a fibre illustrating the Feret diameters (black double-headed arrows) in the 0°, 45°, 90° and 135° positions. It can be seen that the maximum Feret diameter is the F_{45} whilst the minimum is represented by F_{135}. If the tips of all the arrows are joined, an area (the equivalent Feret 'octagon') is obtained which represents a rough measure of the area of space occupied by the particle.

magnification of the optical imaging system in use, the system must therefore be calibrated before use. The intercept measurement is the name given to the sum of the intersections between the horizontal scan lines leaving a detected feature and thus represents a measure of the vertical height of that feature.

If a function computer is fitted to the system then in addition to area, perimeter and the intercept (called 'vertical projection' when obtained from the function computer), it is also possible to obtain horizontal projection and the vertical and horizontal Feret diameter; the Feret diameter (Fig. 4) may be regarded as the diameter which would be obtained if a pair of calipers were used to span the object, and thus takes no account of re-entrancy, unlike the similar measurement termed 'projection'. All of the data derived from a function computer may be subjected to sizing criteria if passed through a classifier/collector.

If a densitometer module is included in the system then the total integrated optical density of a field or feature may be directly obtained in arbitrary figures which can easily be converted to give the integrated density in absolute optical units. By placing the densitometer under programmer control, either the integrated density, the area or the perimeter at each of up to thirty-two pre-programmed density steps may be obtained so that area/density plots may be drawn out. Such techniques of multi-thresholding form valuable tools and are discussed by Bond (1973) who gives selected biological applications.

When the CFFI module is used with a function computer it is also possible to derive the values for the 45° and 135° Feret diameters (Fig. 4). With the addition of these diagonal Feret measurements it now becomes possible to compute the 'equivalent Feret octagon' area which is a measure of the 'roundness' of the space occupied by a particle. In biological work the four Feret diameters may be used to approximate the volume of an ellipsoidal object such as a nucleus in a thick section or a smear using the formula

$$V = \pi/48\,(L+B)^3 \quad \text{(Palkovits \& Fischer, 1968)}$$

where V is nuclear volume, L the maximum Feret diameter and B the minimum Feret diameter. If one considers a straight fibre, it is obvious that the maximum Feret diameter will give a good approximation of the fibre's length. The maximum error due to orientation in this measurement will, as the feature is rotated relative to the direction of the scan line, not exceed *c.* 6 %, as this dimension varies according to a cosine relationship with the angular rotation. The minimum Feret diameter, however, for a similar feature does not give a good approximation to the width. Similarly, it should be noted that as the maximum Feret is the maximum projected length there will be a serious underestimation of the length if the fibre is bent or curly. In order to overcome such problems it is necessary to use more complex parameters derived from the basic measurements.

Derived measures and shape descriptors

Many different derived measurements may be calculated from the basic QTM data; some of those in common use are listed in Table 2. It is often not easy to decide which shape descriptors will give the best chatacterisation of objects without some considerable trial. This is one good argument for taking a large number of measured parameters from each feature onto paper tape so that later those descriptors which experience has shown to be the best for that particular problem may be calculated without the need for an extensive series of remeasurements. This technique is particularly to be commended if the specimens are complex and extensive image editing has to be carried out on each field; it does, however, have the disadvantage of slowing down the measurements to the speed of the slowest unit (usually the tape punch).

In many cases there is little biological information to be gained from a calculation of the shape of a feature; the real worth of such procedures is to allow the QTM to separate and discriminate between data from relevant and irrelevant structures. An example, from a study of cervical carcinoma cells being carried out by the author, is the use of the circularity function A/P^2 to separate data derived from polymorphonuclear leucocyte nuclei (which have a low value for this function) from those derived from the nuclei of cells of the cervix which have much higher A/P^2 ratios. In most preparations from patients with mild or severe dysplasia or with an invasive squamous cell carcinoma there tends to be more or less severe contamination of the preparations with leucocytes and without some form of discrimination, analysis is very difficult. Similar types of shape descriptors are used in the analysis of samples of airborne dust in order to quantify the amount of airborne asbestos fibres (Fraser, 1975).

Requirements for successful application of the QTM 720 to biological problems

Even with the best and most modern automatic image analysis equipment it is still possible to expend a large amount of time and energy and yet end up with meaningless results. Several factors must be considered carefully when designing an investigation involving automatic image analysis in order to obtain valid and biologically significant results.

One of the most important points is to formulate the project in terms of automatic image analysis. This requires careful consideration of the modules available in the particular instrument to be used, an assessment of which parameters are the most suitable for measurement, whether the problem is to be approached on a field or feature-specific basis and so on. Again, the preparation of the specimens is of importance and at the outset one must assess whether there is any significant degree of anisotropy in the objects to be measured and whether they show any marked degree of re-entrance; the answer to this latter question, for example, will govern whether Feret diameters or projections are used. If marked irregularity of shape exists or if there is a marked degree of ellipticity, then the advantages of area sizing rather than chord sizing become apparent. Any histogram of size distribution based upon area sizing is obviously independent of the scan direction and of the orientation of the object whereas diametral sizing is not. The calculation facilities now available on most QTM systems allow a notional diameter to be obtained easily and rapidly if certain assumptions about the shape of the objects are made. A further advantage of area sizing rather than diametral sizing for approximately circular features is that the function r^2 varies faster than $2r$ and hence the sensitivity of the measurement is increased.

In addition to deciding the correct parameters for measurement and the most suitable shape discriminators or derived parameters for calculation, care must be taken to ensure that the resolution obtained on the QTM screen is sufficient for the purpose in hand. Instrumental resolution is based on the picture point, the value of which is obviously going to vary according to the optics of the imaging system in use. For features having linear dimensions of only a few picture points the detected image may, for technical reasons (see Wadlow *et al.*, 1972) be smaller than the observed image, so that the accuracy of the measurements is markedly reduced. In such circumstances if the smaller features are of interest the magnification at which the image is presented to the analyser should be increased.

The accuracy of measurements made on the QTM is dependent upon the scanning, imaging and detection processes. The resolution, geometrical accuracy and uniformity of the scanner are carefully controlled by the makers; the size of the feature compared with the field of view has been mentioned above. The setting of the detection level, however, remains as one of the chief variables under the control of the operator and, as is well known, is one of the chief sources of inaccuracy remaining (Fisher, 1969). With the newer detection modules much of the uncertainty has been removed from this process, but with the standard detector difficulties still remain. Procedures have been developed to reduce error from this source (see for example Wilkinson, Ellis & Callaway (1974) for one such, relevant to the size measurement of latex particles) but this must still be the chief source of error to be controlled at the outset of any Quantimet study.

The number of fields or features which must be measured to obtain statistically valid data is important; if field data such as the number of features per unit area are being recorded, then the number of fields to be scanned is obviously governed by the homogeneity of the specimen. With an on-line calculator it is easy to monitor the data as acquired and continue measurements until some statistical test (e.g. the standard deviation of the mean) reaches an acceptable value. If these facilities are not available then it is generally necessary to carry out a pilot experiment to determine how many features must be measured.

In biological work there is a great need to consider very carefully the nature of the preparative techniques applied to the specimen. Particular care must be taken to keep any fixation, dehydration and embedding procedures as constant as possible in order to avoid the very considerable volume changes, especially shrinkage, which can occur. Experiments in the author's laboratory at Oxford carried out by C. N. Martyn (in preparation) have shown that with brain tissue such variation may still be considerable even when the processing schedules were apparently identical. Much shrinkage could be avoided by substituting an ester wax with a melting point of 38 °C for the standard paraffin wax used in histology. Two other points of importance are that if measurements of, say, nuclear size, are made on sectioned material, then due consideration must be given to the choice of section thickness in order to avoid errors caused by the so-called 'Holmes effect'. This is now clearly recognised in all stereological procedures and a detailed treatment of the effects resulting from this cause on projected nuclear size is to be found in the paper by Weibel (1967). The errors due to the Holmes effect may for practical purposes be minimised if sections of over 10 μm thickness are used and if a sufficiently large number of nuclei is measured. It is, however, worth considering at the outset of such an investigation whether the material could be processed in a way which avoids sectioning. The use of smears or suspensions of isolated or exfoliated cells for measurements of projected area or nuclear volume prevents any measurements of partly-sectioned nuclei being included in the data.

A final point for careful consideration is the method of staining employed for the material. It is desirable to have the maximum contrast in those areas which have to be detected by the QTM 720 with a minimum of stain in the background. If nuclear density measurements are contemplated this may usually be achieved by the use of the Feulgen technique. Here there are no doubts about the chemical specificity of the reaction for the DNA in the nuclei; at the same time the contrast is easily brought to an acceptable level by the use of monochromatic light at 544 nm. It is possible to use the Schiff-type reagents described by Kasten (1958) if the initial contrast of the basic fuchsin of the Feulgen reaction is not sufficient. Azure C has recently been used in this way in my laboratory to increase the nuclear contrast for a series of measurements of the volume changes of liver nuclei following bile duct ligation. Similar considerations have led Miller & Alston (1975) to use a new quinolinic phthalocyanin dye for nucleic acids in their study of neurones in the brain. If elastic fibres are to be quantified, it is far better to stain them with the Verhoeff technique (which colours them black) rather than with orcein (which colours them reddish-brown); there is very little loss in specificity but a tremendous gain in the ease with which the fibres may be detected.

Applications of the QTM 720 in biology

As automatic image analysers become more common in biological laboratories, the number of published papers reporting their application has increased rapidly. Jesse (1974) points out that in the three years 1971–4 the number of relevant papers increased by a factor of four compared with the preceding six years. It is not possible in this paper to survey all the applications now reported; some indication of the variety of fields may gain by inspection of the bibliography published in *The Microscope* (1974).

The QTM may be used as a very sophisticated tool for obtaining measurements which could be taken manually but with far less speed and accuracy; an example of this is provided by its use for the measurement of the fascicle length of muscle fibres following joint immobilisation and during growth (Crawford, 1975). More commonly, however, the instrument is used to accumulate measures which could be almost impossible to obtain manually. A topic of great interest is the changes in the nerve fibres of the sciatic nerve of normal mice at different ages and in mice of similar ages kept for long periods on the anti-leprosy drug 'Dapsone' (diamino diphenylsulphone); our work has allowed the preparation of area sizing histograms of both the axons alone and of the total fibres. From the measurements we have calculated the axon notional diameter and compared it with the myelin sheath thickness for each experimental group and each age group in the form of scatter plots.

The project was undertaken as follows. Transverse ultrathin sections of the whole nerve were prepared from animals at 3, 6, 9, 12, 15, 18, 21 and 24 months of age and electron micrographs taken. Initially, owing to contrast problems and the difficulties of detecting the relevant areas on the QTM, tracings of the axons and their myelin sheaths were prepared using indian ink and measurements done on these. Recently, with the addition of an image editor to our system, it has proved possible to quantify the electron micrograph negatives by directly imaging them on the scanner using the epidiascope attachment. In each case the sections were from the same area in the proximal region of the main sciatic branch of the nerve and at least 1000 fibres per specimen were measured, with three separate animals of each age group being examined. Data from each fibre were obtained by the use of the CFFI module and taken onto paper tape. Calculation enabled the geometrical mean sectional area of the axons and fibres to be obtained together with their standard deviations, but much more valuable information was gained from the plotting of the normalised histograms of fibre and axon area sizing and from the scatter plot of the notional diameter of the axon against the thickness of the myelin sheath.

Some of our data are presented as Fig. 5, from which it may be seen that although there are some changes in the pattern between the normal and drug-treated animal up to 12 months of age, a dramatic difference appears when the 18-month size distributions are compared. At this age the drug-treated

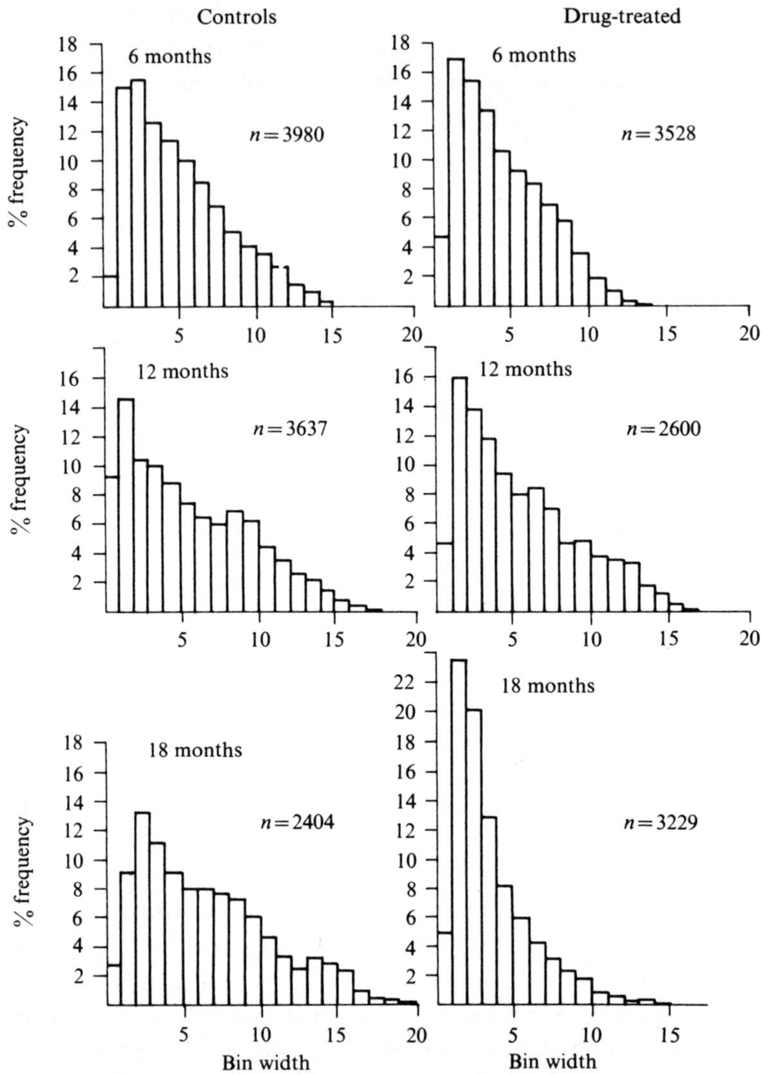

Fig. 5. Histograms of the distributions of nerve fibre area in the sciatic nerve of mice aged 6, 12, and 18 months both normal and on long-term treatment with 'Dapsone'. Each bin width = 20 μm^2.

animals show a very pronounced reversion to a pattern reminiscent of the juvenile, a finding which we interpret as due to the presence of large numbers of regenerated fibres. The full data and a discussion of their significance will be published in a paper by Stanmore, Bradbury & Weddell now in preparation.

The QTM has also been applied in our laboratory to the measurement of biopsy material obtained from the normal and abnormal cervix uteri. The biopsies of material were obtained in the hospital and processed in the laboratory for the preparation of electron micrographs. These were subsequently examined using the epidiascope of the QTM 720 in order to derive numerical information on parameters which would help us distinguish objectively between normal, the several varying degrees of severity of dysplasia (as classified by the subjective criteria of the pathologist), carcinoma *in situ* and overt invasive squamous cell carcinoma. We found, for example, that it is possible to obtain estimates of the percentage area of the tissue occupied by the extracellular matrix. When compared with normal tissues, the specimens obtained from cases of cervical carcinoma showed marked and significant reduction in the area of extracellular matrix (Wiernik *et al.*, 1973). From electron micrographs of the biopsies it also proved possible to use the image editor to measure the percentage of the projected cell membrane which is occupied by desmosomes. Fig. 6(*a*) shows a micrograph on the screen of the QTM 720 with the desmosomal specialisations of the cell membrane indicated by the bright 'line' display of the image editor; the total length of this line in picture points is displayed at the top of the screen. Fig. 6(*b*) shows the second stage in the measurement, when the total extent of the cell membrane has been traced with the image editor light pen. Again the length is measured in picture points and calculation of the percentage occupied by the desmosomes is now simple. In practice a large number of micrographs were sampled, data accumulated in the calculator and a mean value obtained; repeating the procedure on biopsies from normal specimens allowed comparisons between the normal and abnormal. We have shown (Wiernik *et al.*, 1973) that the desmosome values for abnormal cells are much smaller than for the normals, a finding which may be related to the tendency of the carcinoma cells to lose contact with each other, invade underlying tissues and form metastases. Recently we have extended our studies to the examination of the nuclei of dysplastic cells. These were obtained at colposcopic examination by placing a slice of millipore filter paper on the area of the cervix showing the lesion. When the paper is removed, the superficial cells adhere to it and may be fixed in alcohol and stained by the standard Feulgen procedure. From such preparations it is possible to measure the optical density of the nuclei with the densitometer and at the same time obtain figures for the area, perimeter and four Feret diameters of each nucleus. The image editor allows touching cells, debris and leucocytes to be easily excluded from the measurements; in good preparations the nuclei of up to 500 cells may be measured in a morning. From the optical density measurements it is possible to produce histograms which show whether the dysplastic cells are significantly polyploid,

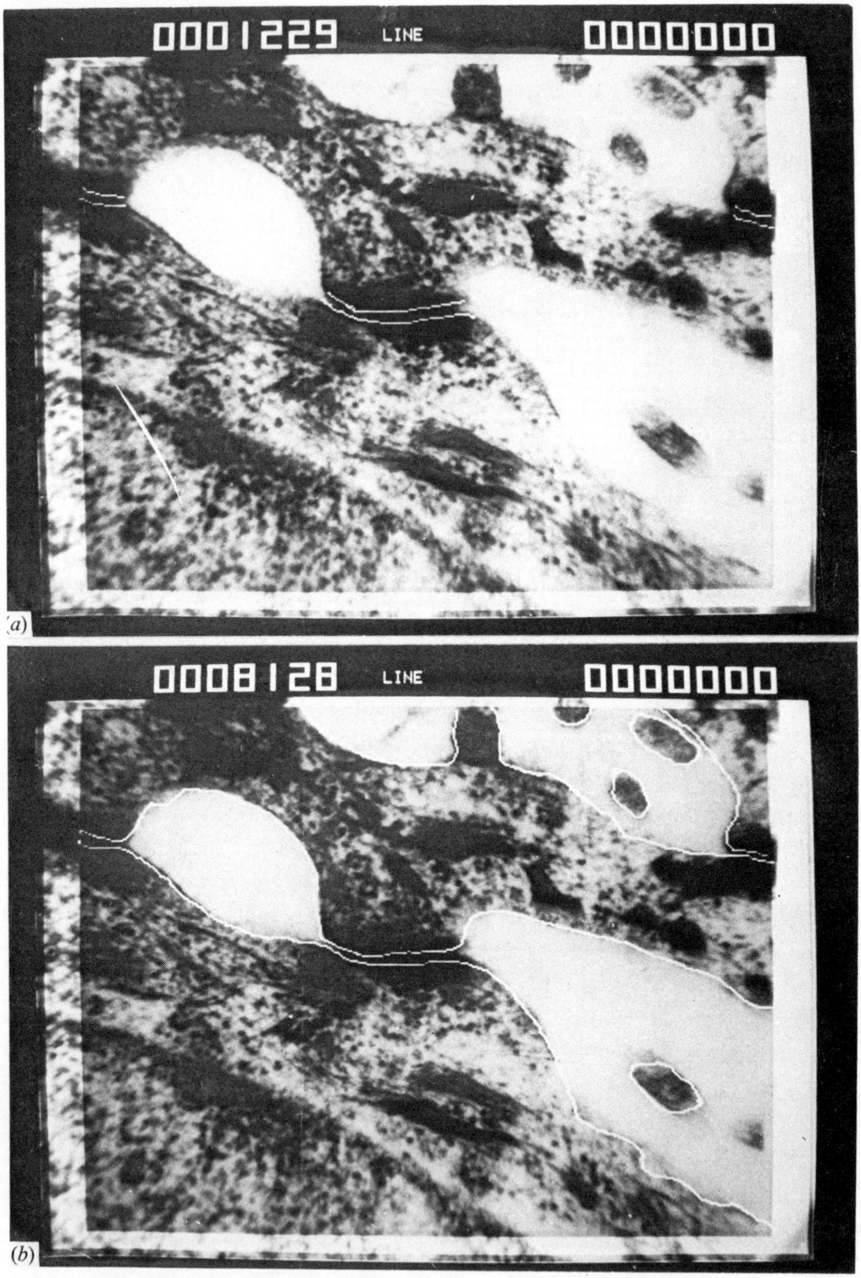

Fig. 6. (*a*) An electron micrograph of a biopsy of cervical epithelia, as seen on the screen of the QTM 720. The image editor, in 'line' mode has been used to select the desmosomal specialisations on the cell membranes, the length of the measured desmosomes being shown as 1229 picture points. (*b*) The same micrograph with the total projection of cell membrane outlined. The length is 8128 picture points, so that in this field 15 % of the total length is occupied by the desmosomes.

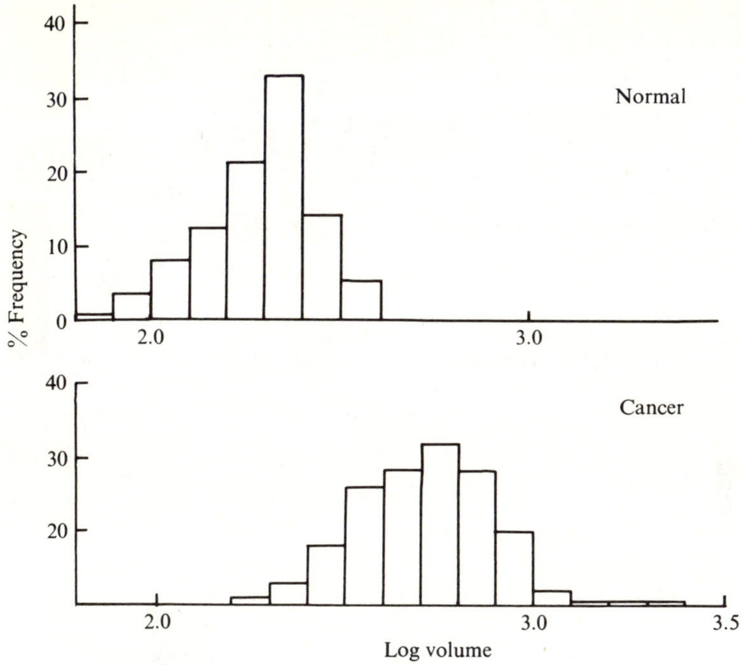

Fig. 7. Histograms to show the log volume distributions of nuclei of normal and carcinomatous cells from the cervix uteri. Note the large increase in the volume of the abnormal cell nuclei.

whilst the Feret diameters enable us to produce histograms of the nuclear log volume. Fig. 7 clearly shows the large increase in the volume of the nuclei of cervical cells in a case of squamous cell carcinoma. It is hoped with such studies to produce numerical data which will clarify the rather subjective descriptions of abnormal conditions at present in use by cytopathologists.

One final example of the role of the QTM 720 in providing objective numerical data is work carried out on the analysis of sonogram recordings of chaffinch song (Pickstock, Krebs & Bradbury, in preparation). Sonograms of songs of the chaffinch (*Fringilla coelebs*) were prepared from recordings made in Oxford and in North Wales. The end phrase of the sonogram (labelled C in Fig. 8) was analysed for 148 birds and the area, perimeter and vertical projection measurements taken for each. It was obvious by visual inspection that the end phrases could be classified into four main patterns (Fig. 9) but when the measurements of A/P^2 and VP were compared, it was found that there was an extremely strong negative correlation between these two, as may be seen from Table 3. Statistical testing of the data from all the

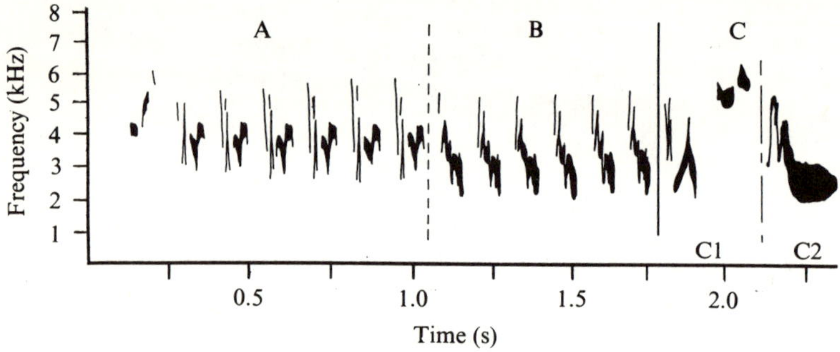

Fig. 8. A sonogram of the chaffinch song showing the relationship of frequency (kHz) to time (s). The end phrase is labelled C, to the right of the line at 1.8 s.

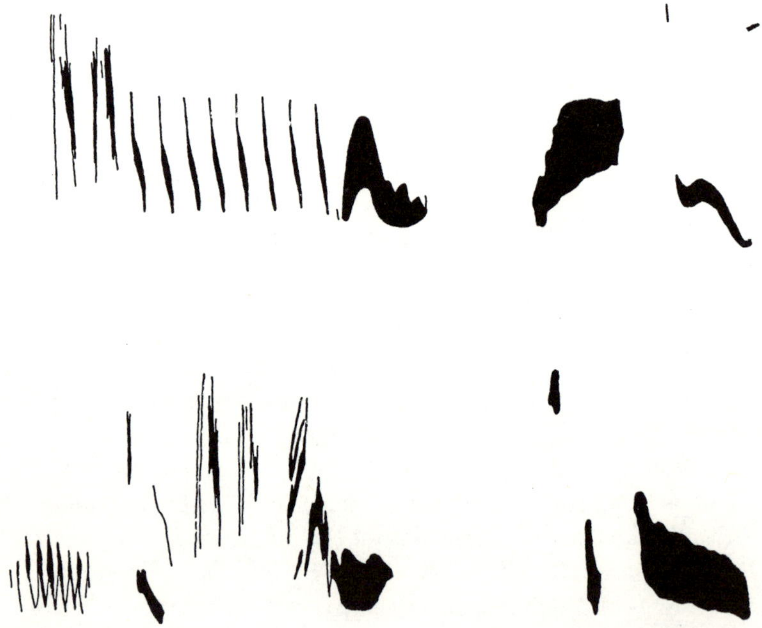

Fig. 9. Four main end phrase patterns of the chaffinch song. These show a high negative correlation between their values of A/P² and vertical projection.

Table 3. *Numerical values for A/P^2 and vertical projection from four selected chaffinch sonogram end phrases*

Bird number	Shape (A/P^2)	Vertical projection
P7a	6×10^{-4}	1652
PN 11b	3.8×10^{-4}	2057
PN 4b	2×10^{-2}	227
D 15b	1.6×10^{-2}	262

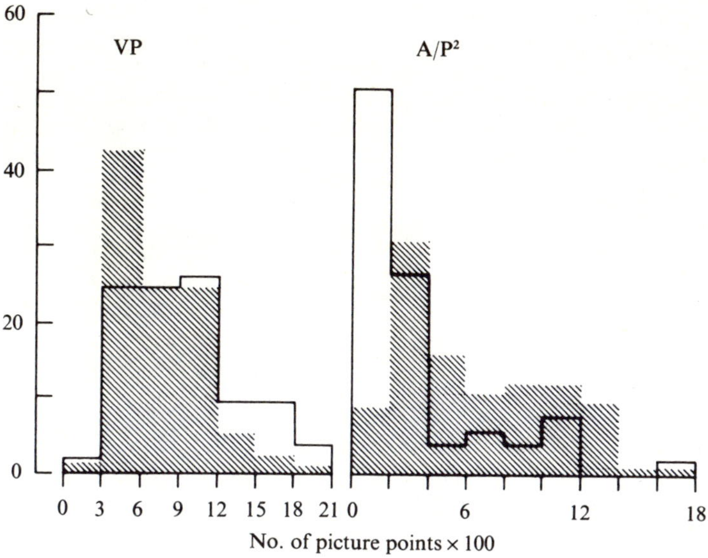

Fig. 10. Histograms of values of the shape coefficient (A/P^2) and vertical projection (VP) showing the marked differences between chaffinches recorded in Oxford (shaded areas) and in North Wales (open lines).

ninety-five birds recorded in Oxford showed a very high Spearman's rank correlation coefficient between the two measurements of -0.963, whilst for fifty-three end phrases recorded in North Wales the correlation was -0.888; both of these proved significant at the 1 % level. The comparison of the distributions of shape and projection measurements between Welsh and Oxford birds may be clearly demonstrated graphically and is illustrated in Fig. 10. Such studies may prove of value to the ornithologists in their studies of the learning processes involved in the transmission of bird song patterns.

Clearly the application of automatic image analysis methods has a tremendous potential in biology, not only in the obvious fields of cell biology and comparative pathology but also in allowing quantitative and statistical techniques to be used in many biological fields which are not immediately obvious as allowing this approach.

References

Bond, C. P. (1973). Television microdensitometry measurements in biological image analysis. *Proceedings of the Imanco Symposium, Frankfurt 1973*, pp. 33–41.

Crawford, G. N. C. (1975). The growth of muscle following tenotomy. *Journal of Anatomy*, **120**, 581–93.

Fisher, C. (1969). The Quantimet: setting the threshold and the correction of off-set threshold error. *Praktische Metallographie*, **6**, 659–72.

Fisher, C. (1971). The new Quantimet 720. *Microscope*, **19**, 1–20.

Fraser, P. V. (1975). Asbestos ... an exercise in pattern recognition. Paper published privately by Image Analysing Computer Ltd (now Cambridge Instruments, Melbourn, nr Royston, Herts).

Jesse, A. (1974). Bibliography on automatic image analysis. *Microscope*, **22**, 89–115.

Kasten, F. H. (1958). Additional Schiff-type reagents for use in cytochemistry. *Stain Technology*, **33**, 39–45.

Miller, A. K. H. & Alston, R. L. (1975). The effects of ageing on the brain. *Quantimet*, no. 2. Cambridge Instruments, Melbourn, nr Royston, Herts.

Palkovits, M. & Fischer, J. (1968). *Karyometric Investigations*. Akademiai Kiado, Budapest.

Wadlow, D. E., Hopkins, B. M., Gardner, G. M. & Fisher, C. (1972). New detection systems for the Quantimet 720. *Microscope*, **20**, 183–202.

Weibel, E. R. (1967). Stereological principles for morphometry in electron microscopic cytology. *International Review of Cytology*, **26**, 235–302.

Wiernik, G., Bradbury, S., Plant, M., Cowdell, R. H. & Williams, E. A. (1973). A quantitative comparison between normal and carcinomatous squamous epithelia of the uterine cervix. *British Journal of Cancer*, **28**, 488–99.

Wilkinson, M. C., Ellis, R. & Callaway, S. (1974). The evaluation of the use of the Quantimet 720 in the determination of the size distributions of monodisperse lattices. *Microscope*, **22**, 229–45.

D. J. GOLDSTEIN

Integrating microdensitometry

General principles

When a beam of light passes through a solid, liquid or gaseous medium some of the light is usually absorbed and converted into other forms of energy, for example heat. The interaction of matter with visible light, which has a wavelength between about 400 and 700 nm, mainly involves the outer electron shells of the atoms concerned, and is a function of the chemical composition and to a lesser extent the environment and physical state of the material. The absorption of light is in general wavelength-dependent, making the material appear coloured. According to the Bouguer–Beer absorption law, if a parallel beam of light of a given wavelength passes through a homogeneous solution the absorbance (defined as the logarithm of the reciprocal of the fraction of the light finally emerging from the solution) is directly proportional to the pathlength through the solution and to the concentration of the light-absorbing solute. The proportionality constant characteristic of a particular wavelength and substance is called the absorptivity. Where the law holds, the absorbance is proportional to the mass of solute per unit area of solution as seen from the direction in which the light is travelling, i.e. to the mass of solute contained in a column of solution of unit cross-sectional area.

The intensity of light being relatively easily measured, absorptiometric (densitometric) methods of quantitatively analysing concentrations or amounts of light-absorbing substances are commonplace in the biological and physical sciences. In principle all that is required (Fig. 1) is a light source, some means of rendering the light adequately monochromatic and parallel, a specimen holder and a light-detector. A typical light-detector is a photomultiplier tube, the electrical output of which is directly proportional to the intensity of light striking the photocathode. By measuring the intensities of light I_t and I_0, respectively with and without the specimen in position, the sample absorbance $E = \log (I_0/I_t)$ is readily obtained.

The extension of densitometry to small, non-uniform objects such as stained cells or tissues requires the addition of a high-aperture, high-resolution microscope to the optical system (Fig. 2). It is immediately apparent that

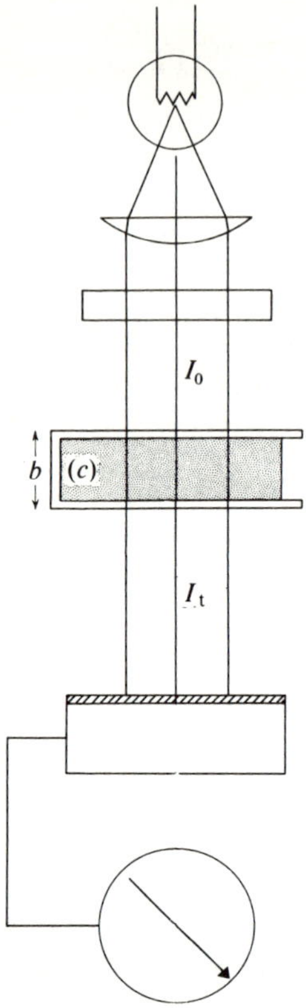

Fig. 1. Schematic representation of a spectrophotometer. Absorbance $E = \log (I_0/I_t) = a \cdot b \cdot c$, where a, b and c are respectively the absorptivity (constant), the pathlength through the solution and the concentration of the solute.

the light passing through the specimen is no longer the parallel beam called for by the Bouguer–Beer law, but perhaps surprisingly this is unimportant in most cases (Walker, 1956). Certain less obvious difficulties are, however, far from negligible, and some of the more significant of these will briefly be described (for more general discussions of microdensitometry see e.g. Mayall & Mendelsohn, 1970*a*, *b*).

'Distribution error' is due essentially to spatial heterogeneity of the specimen. Since (according to the Bouguer–Beer law) the amount of light-absorbing material per unit projected area of specimen is proportional to the absorbance E, the total amount of such material in a discrete, non-uniform microscopic specimen is proportional to the total projected area multiplied by the mean of the individual absorbances of each separate point of the specimen.

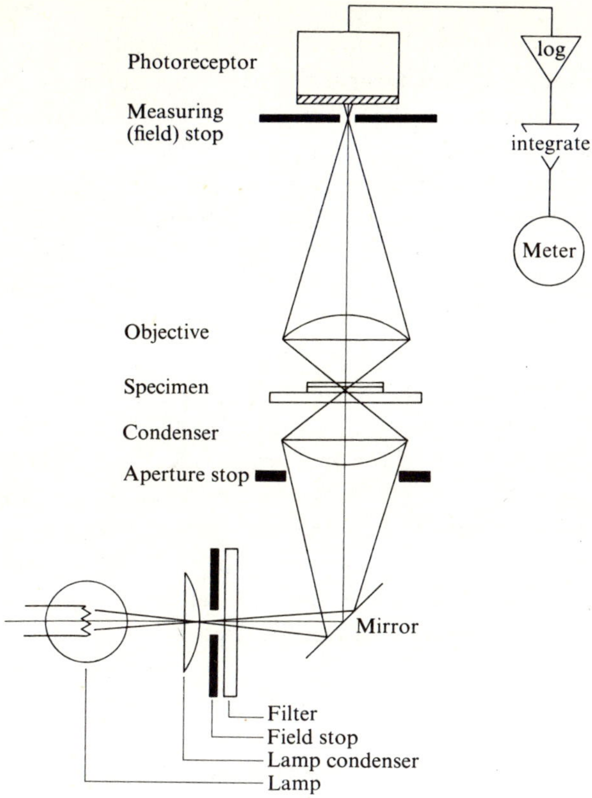

Fig. 2. Schematic representation of a microdensitometer. See text for details. (The scanning mechanism for moving either the object or its image is not indicated.)

To estimate this value, known as the 'integrated absorbance', it is not sufficient merely to multiply the measured projected specimen area by an absorbance value calculated from the measured average transmittance of the whole microscopic field or object. A result obtained in this way will always be lower than the true value: those who are mathematically minded may note that the logarithm of a mean is not equivalent to the mean of the logarithms of the individual values. It is possible partially to correct for distribution error by measuring the mean transmittance of the specimen at two different wavelengths, and applying appropriate formulae (Mendelsohn, 1966), but the most direct and generally satisfactory approach to the elimination of distribution error is the use of a scanning and integrating microdensitometer.

In such an instrument a suitably situated field stop limits the specimen measured at any given moment to an area small enough to be effectively homogeneous. Light passing through this area is received by a photodetector, the output of which is the electrical analogue of the transmittance. An electronic 'analogue computer' (i.e. a logarithmic amplifier) automatically converts the transmittance signal into the corresponding analogue absorbance value. The specimen is scanned in a systematic pattern ('raster') by moving the measuring spot relative to the object, and by automatically adding ('integrating') the instantaneous absorbances registered during the course of the scan the required 'integrated absorbance' of the specimen is obtained more or less free from distribution error. The spatial resolution of a scanning system is, however, only 'diffraction limited' (i.e. as good as is theoretically obtainable with the particular objective aperture and wavelength of light employed) if the diameter of the measuring spot is less than about half the theoretical limit of resolution (Welford, 1972). Using green light and a good oil-immersion objective the diameter of the measuring spot should ideally be not greater than about 0.125 μm. This is seldom achieved in practice, since with very small measuring spots so little light reaches the photodetector that the signal-to-noise ratio of the whole system may be too low for reproducible results to be obtained. The measuring spot is therefore usually somewhat larger than the finest theoretically resolvable specimen detail, but the resulting second-order or residual distribution error is often negligible except in the most exact work. It can be shown that to a first approximation the deficit in the apparent integrated absorbance due to residual distribution error is proportional to the diameter of the measuring spot. With some types of microdensitometer it is possible to assess, and if necessary correct, residual distribution error by measuring the specimen using measuring spots of different diameters, and extrapolating the apparent integrated absorbances thus obtained to estimate the value corresponding to zero spot diameter (Goldstein, 1971).

Errors in integrating microdensitometry which are theoretically closely related to distribution error occur if the specimen is incorrectly focused, or if the specimen is thicker (parallel to the optical axis of the microscope) than the depth of focus of the objective. The use of thick sections should therefore be avoided, and well flattened smear or squash preparations should be employed whenever possible.

Stray light, or glare, is an important source of systematic microdensitometric error occurring with specimens incompletely or non-uniformly filling the illuminated microscopical field. Due to unavoidable reflections and light-scattering in the optical system, a proportion of the light coming from a given point in the object is inevitably distributed over most or all of the image plane. Some light from bright, specimen-free background areas therefore finds its way into the image of dark parts of the object, which consequently appear to have higher transmittances (lower absorbances) than is actually the case. The percentage error in the apparent absorbance increases with the true absorbance, so systematic errors occur in the comparison of lightly and darkly stained specimens. Glare error can be minimised by close attention to the cleanliness of the optics and by restricting the area of specimen illuminated. The amount of glare remaining in the system can be estimated by measuring the apparent transmittance of an opaque object, such as a carbon particle, having about the same dimensions as the biological specimen of interest. With certain microdensitometers it is possible to compensate for glare, if this is found to be necessary, by electronic off-setting of the current coming from the photomultiplier tube (Goldstein, 1970).

In any form of densitometry it is essential to employ light of adequate spectral purity. Most microdensitometers employ a light source such as a tungsten lamp, which emits light having a continuous spectrum. The desired light must then be isolated by the use of interchangeable spectral filters, or more conveniently and flexibly by a grating or prism monochromator or a continuous-spectrum interference filter. Methods have been described (Goldstein, 1975) to assess and if necessary correct various microdensitometric errors connected with the lamp and monochromator system, including incorrect setting of the wavelength, variation of the wavelength across the microscopical field, and the use of light with too wide a spectral bandwidth. Substantial errors can occur from such causes, but should be avoidable with modern equipment. This is also true of less common instrumental sources of microdensitometric error, such as the absorbance of the specimen exceeding the measuring range of the instrument.

Instrumentation

Several types of scanning and integrating mirodensitometer are commercially obtainable in this country, and new models appear from time to time (Table 1). All are capable of measuring the integrated absorbance of an isolated object such as a Feulgen-stained nucleus, and can also scan along a single raster line and with the aid of an oscilloscope, pen-recorder or similar ancillary device record an 'absorbance profile' across a specimen. An area-measurement facility usually permits the measurement of specimen areas having absorbances above or between arbitrarily set thresholds. With the scanning mechanism inactivated, 'spot' readings of absorbance or transmittance can easily be made with some microdensitometers. Most instruments have additional capabilities of one sort or another, and some can be built up into systems of considerable complexity. The choice facing a prospective

Table 1. *Comparison of commercial microdensitometers*

Instrument	Type of scan	Scan time (s)	Spot diam. (μm)	Line spacing (μm)	Maximum *E*	Monochromator type and min. bandwidth	Typical precision claimed	Diam. of field illuminated (μm)	Typical glare value	Provision to: offset glare	Provision to: vary spot diameter	Provision to: make spot readings	Cost
Barr & Stroud GN5	Image	4	0·25	0·25	2·5	Filters 18 nm	< 1 %	100	—	No	No	No	£5 775
Imanco Quantimet 720	TV	0·1	0·1	0·1	2·52	Filters	< 2 %	70	—	Yes	No	No	£11 950
Leitz MPV	Object	3	0·5	0·5	—	Grating 6 nm	0·5 %	1·5	—	Yes	Yes	Yes	*c.* £18 400
Vickers M85	Flying-spot	3	0·25	0·20	1·3	Prism 5 nm	0·25 to 0·5 %	100	2–3·5 %	Yes	Yes	Yes	£6 292
Zeiss MPM 01	Object	35	0·5	0·5	2·37	Filters 6 nm	1·8 %	1·5	—	Yes	Yes	Yes	*c.* £17 000

buyer or user is therefore somewhat bewildering, and although certain microdensitometers have recently been described in some detail by Altman (1975), a brief discussion of currently available models may be useful.

In order to have some baseline for comparison the instruments considered are the simplest and cheapest of their respective types capable of integrating microdensitometry, but a direct comparison of costs may nevertheless be unfair to the more complex, versatile and expensive instruments. It should also be noted that no impartial, independent experimental evaluation of the various microdensitometers is available; the data quoted are therefore those supplied by the respective manufacturers, and may not be directly comparable.

Integrating microdensitometers may be classified according to the manner in which the scanning is executed. Thus in moving-object instruments (e.g. Zeiss, Leitz) the measuring spot is stationary and the object itself is moved by a motor-driven stage. In the Barr & Stroud microdensitometer an enlarged image of the stationary specimen is scanned by a mechanical system, while in the Vickers flying-spot instrument the optically reduced image of an illuminated measuring aperture is moved in the plane of the stationary object by a pair of oscillating mirrors. In television systems the microscopical image received by a television camera is analysed by purely electronic means. All types of scanning system are in principle equivalent with respect to the resolution attainable, but in practice each system has certain advantages and disadvantages.

In moving-object microdensitometers the field illuminated needs only to be slightly larger than the area actually measured at any given moment. This is helpful in minimising glare. Because the same central part of the microscopical field is always utilised, the system is unaffected by off-axis optical aberrations, and it is also possible to measure specimens which are larger than the field of view of the objective. Since the scanning mechanism operates on the microscope stage, the microscope tube and substage can be standard, interchangeable components, and substitution of alternative modules permits the application of a wide variety of microscopical techniques such as incident illumination, fluorescence and ultraviolet microscopy. Automatic wavelength scanning to obtain a transmission or absorbance spectrum of a given part of a stationary specimen is possible with some instruments. If the microdensitometer output is fed into an external computer, microscopical images can be stored in digital form and analysed by more or less sophisticated computer techniques. On the debit side, these elaborate microscope systems are correspondingly expensive and complicated to operate. Motor-driven stages are relatively slow, and in order to complete a scan (raster) in a reasonable time the measuring spot and the spacing of the raster lines are usually fairly large;

as a consequence the effective resolution of such systems is not particularly good under normal conditions of use, and scanning and integrating measurements may be liable to residual distribution error. It may also be quite difficult to limit the measured area to a small or irregularly shaped object, since in general the operator has control only over the x and y dimensions of a rectangular raster.

The image-scanning Barr & Stroud GN5 and the flying-spot Vickers M85 microdensitometers are conveniently considered together. Both instruments are (relatively) simple and inexpensive, and are built specifically for high-speed integrating microdensitometry. The Barr & Stroud GN5 is slightly cheaper and more compact than the Vickers M85 and is probably somewhat simpler to operate. It is not, however, possible to vary the measuring-spot size of the Barr & Stroud without changing objectives, so that residual distribution error cannot be assessed or corrected. No facility is provided for the electronic off-setting of glare, and only a limited range of wavelengths is available using interchangeable spectral filters. The 'masking' system for defining the area to be measured is considerably less flexible than that provided on the Vickers M85, which in all these respects appears to be the more versatile and powerful instrument. The Vickers microdensitometer possesses in addition the unique feature of being convertible into a scanning and integrating micro-interferometer capable of measuring the dry mass of unstained cells or tissues (see Goldstein (scanning microinterferometry) this volume).

The Quantimet (Imanco) television microscope can be coupled to a computer for both control and analytical purposes, but is intended primarily for extremely rapid image analysis using 'hard-wired' logic modules. These permit the measurement of a wide variety of specimen parameters, on the basis of which automatic classification and counting can be carried out (see Bradbury, this volume). The more elaborate installations can even focus automatically, and search a microscopic preparation for suitable objects (e.g. metaphase chromosome plates) for detailed study. The resolution of the television camera is such that the microdensitometer version should be essentially diffraction-limited and free from distribution error. Glare can be off-set manually, or can be measured and compensated automatically. The Leitz microscope body usually employed will accept various accessories, including a monochromator instead of the interchangeable spectral filters specified in Table 1. All the requirements for high-accuracy integrating microdensitometry therefore appear to be met by the instrument, but even the cheapest microdensitometer version of the Quantimet is considerably more complex and expensive than the specialised Barr & Stroud and Vickers instruments. The remarkable speed of television scanning is probably not a major advantage for most microdensitometry, since there appears to be little to be gained from a scan taking only a fraction of a second if considerably longer than this is required to locate a suitable specimen and adjust the apparatus. The real strength of the Quantimet lies in its image analysis capability and versatility, and probably few are purchased solely for use as a microdensitometer.

In comparing microdensitometers it should be remembered that a relatively rapid instrument is able to make more measurements in a given time, and hence attain a greater effective precision, than a slower machine of otherwise similar specification. Precision is not, however, the main problem with modern instruments, all of which ought to be capable of yielding results which are sufficiently reproducible for most biological purposes. Accuracy, or the closeness with which the apparent result approaches the true value, is much more difficult to assess than mere precision (reproducibility), but several commercial integrating microdensitometers possess facilities which enable all currently known major sources of instrumental error to be eliminated or corrected. In choosing an instrument with a particular application in mind, many factors must be taken into account in addition to precision and accuracy. The rapidity and ease of use of the equipment on the one hand, and its versatility and possible use in other investigations on the other, are clearly matters of importance, while the financial cost is often the deciding factor. Many different instruments are now in use in United Kingdom laboratories, and it is highly desirable for a prospective user of a particular make of microdensitometer to carry out a pilot investigation with similar equipment to ascertain its suitability for the work contemplated. In any event impartial advice from experienced, existing users should certainly be sought before buying any apparatus.

Biological applications of integrating microdensitometry

Contrary to an impression which might be gained from some current literature, the generation of numerical data is not a valid end in itself. The purchase and use of relatively expensive and complicated microdensitometric apparatus therefore requires justification, and some applications of integrating microdensitometry will be discussed to illustrate the scope of the techniques.

Suitably stained smear or squash preparations of whole, isolated cells or nuclei have been widely used for microdensitometry. Since the whole object is present on the slide statistical problems associated with sectioning do not arise, and the specimen is usually sufficiently flattened for focusing difficulties to be avoided. Whole mounts of some relatively flat and extended tissue cultures have also been successfully studied by microdensitometry, and sections can be used satisfactorily for some types of work provided care is taken to use suitable controls. Sections have the major advantage of retaining tissue structure and cell relationships, without possible biasing of results by differential loss during tissue dispersion and cell isolating procedures. Random variation of section thickness can be a real problem, but has perhaps been overemphasised by some workers; it can usually be allowed for by the simple expedient of measuring several different sections. Uncertainty about absolute section thickness can sometimes be overcome by interference microscopy, or by the use of comparative measurements. Thus the percentage change in (say) the glycogen content of liver due to some experimental procedure can be ascertained using sections of unknown absolute thickness, provided experimental and control tissues are identically and preferably simultaneously prepared for examination.

In addition to applications involving living or fixed cells and tissues, microdensitometry has been used successfully in the measurement of minute quantities of substances such as phosphorus, nucleotides and various enzymes, using conventional in-vitro chemical reactions (see e.g. Thorell, 1958). In recent years there appears to have been less interest in this type of work, possibly because of the development of other highly sensitive assay methods such as scintillation counting and fluorescence measurement (see Ploem, this volume). Modern integrating microdensitometers can, however, measure rapidly and precisely the total amount of coloured reaction product in a microdroplet even of unknown geometry and volume, and it is conceivable that microdensitometric assay methods may return to favour. Techniques have in fact recently been described (Jongkind, Reuser & Galjaard, 1975) for the measurement of the enzyme content of single, isolated cells using microdensitometric and microfluorescence measurements on submicrolitre volumes of fluid, and it should be possible to adapt such methods for use in a wide variety of problems.

Irrespective of whether they occur in smears, sections or solution, substances measurable by microdensitometry share the property of being light-absorbing. For the purposes of the present discussion we may somewhat arbitrarily classify such substances as inherently coloured, empirically stained, specifically stained, or opaque. These groups will be dealt with in turn.

Many naturally coloured substances in animal and plant tissues have been or might be successfully investigated by microdensitometry. Cytochrome *c* was observed microspectrophotometrically in living cells as long ago as 1886 (MacMunn, 1886; Keilin, 1925), and as in the case of verdoperoxidase (Thorell, 1958) might be measurable by integrating microdensitometry especially if concentrated inside the cell by centrifugation. Cytochrome P450 has been measured in unfixed, unstained cryostat sections of liver, and differences demonstrated between control and phenobarbitone-treated animals (Altman, Moore & Chayen, 1975). Photosensitive retinal cone pigments are probably not readily measurable with standard instruments, but methods for studying such pigments with a special low-light-level microspectrophotometer (Liebman & Entine, 1964) might be adaptable to commercially available equipment, the difficulty lying more in the preparation of the specimen than in the actual measurement. Other intrinsically light-absorbing pigments of functional importance in higher organisms include carotene, chlorophyll, melanin, myoglobin and haemoglobin. All of these should be measurable, at least in cells containing relatively high amounts, and haemoglobin has in fact been studied by a number of workers (for references see James & Goldstein, 1974). In the microdensitometry of unstained haemoglobin the *c.* 415-nm absorption band (first described by Soret) is usually used in preference to the less intense green absorption responsible for the familiar red colour of blood. There are naturally no difficulties with staining specificity or stoichiometry in work with unstained haemoglobin, but other problems arise if high microdensitometric accuracy is sought. Thus the chemical state of haemoglobin in air-dried blood smears is not known with certainty, and although data are available for the absorptivities of haemoglobin, oxyhaemoglobin, methaemoglobin and other forms of the substance in solution, it is not clear which if any of these constants is appropriate for microdensitometry. There are also indications that part of the integrated absorbance of an unstained erythrocyte measured at 415 nm may be due to substances other than haemoglobin. Despite these difficulties a good correlation can be found between the mean corpuscular haemoglobin of a given sample of blood as measured by microdensitometry and by conventional haematological procedures; microdensitometry in addition enables variations between individual cells in a given sample to be studied, and can reveal differences, e.g. between normal and pathological human blood.

Thousands of empirical staining methods have been described in a voluminous literature dating back more than a century, and some of the scores or hundreds of procedures still in regular use are indispensable to the practising histologist and cytologist. Highly selective methods exist for the staining e.g. of collagen fibres, elastic fibres, mucins of various sorts, nuclei, different cell types and hosts of other structures in smears or sections, but it is useful to draw a distinction between 'selective' and 'specific' as applied to a staining method. 'Selective' implies that only certain structures or cells are stained, i.e. that a few items are picked out from among the many present. A good example is the classical Golgi method for the metallic impregnation of neurones, which demonstrates only a small proportion of the nerve cells in a given section but shows these up apparently completely. 'Specific' means a good deal more than this, a specific method depending on some known *characteristic* property of the stained object and not on some trivial, accidental or obscure one. In addition the method should be clearly defined, its limitations known and its mechanism understood, at least in broad outline. These criteria do not hold for the majority of traditional staining methods, and it may appear futile to apply sophisticated and precise technology to the measurement of dye uptake of uncertain quantitative and chemical significance. Nevertheless, microdensitometry is at least objective and reproducible, and enables results obtained with different tissues or at different times to be analysed statistically and compared. For example, measurement of the uptake of a selective mucin stain might be an effective way to demonstrate experimental changes in gastrointestinal or respiratory epithelium, and empirical collagen staining could be applied in quantitative work on wound healing. Many other possibilities come readily to mind. Even under circumstances where measurements of integrated absorbance may be difficult or impossible to interpret (e.g. with highly absorbing or opaque objects), the use of an empirical but selective staining or impregnation method may increase the contrast of a particular tissue component sufficiently to permit its quantitation by the area-measuring device incorporated in most scanning microdensitometers.

The specificity of empirical staining methods can sometimes be considerably increased by the use of extraction procedures. For example, Best's carmine method stains glycogen, certain mucins and sometimes other structures, but the use of control sections from which glycogen has been removed by salivary diastase enables the method to be used histochemically. Substances removable from sections by specific enzymic or chemical treatment include RNA, DNA, histones, collagen and hyaluronic acid, while blocking procedures exist for the elimination from specimens *inter alia* of phosphate, carboxyl and amino groups.

This brings us to the subject of specific histochemical and cytochemical staining methods, many of which are available for qualitative use with both the optical microscope and the electron microscope. Relatively few have, however, been applied in quantitative work and not all of these have been studied sufficiently thoroughly for their reliability and validity to be accurately assessed. No general review of the extensive and controversial literature can be attempted here (see however Pearse, 1968, 1972; Wied, 1966; Wied & Bahr, 1970; Ruch, 1972), but a few of the most important methods will be mentioned.

The histochemical method most frequently used in microdensitometry is undoubtedly the Feulgen reaction for DNA. Some details of its mechanism are still obscure, and objections against the quantitative use of the method are raised from time to time, but the procedure appears to be reliable when properly applied. Discrepancies can mostly be attributed either to faulty technique such as incorrect fixation (Böhm, Sprenger, Schluter & Sandritter 1968) or to instrumental errors like glare and residual distribution error (Bedi & Goldstein, 1975).

Other histochemical methods which have been used quantitatively include variants of the Millon reaction for tyrosine and tryptophane, the Sakaguchi reaction for arginine, the Mercury Orange and DDD methods for disulphide and sulphydryl groups, various acid dyes used at low pH for total basic amino acids and at high pH for histones and protamines, the periodic acid-Schiff reaction for carbohydrates and the ninhydrin-Schiff procedure for amino groups. Numerous enzymes can be demonstrated histochemically, and attempts at quantitation have been made, e.g. for acid phosphatase (Bácsy & Rappay, 1971; Friede & Knoller, 1965), alkaline phosphatase (Friede & Knoller, 1965; Cabrini, Frasch & Itoiz, 1975), succinic dehydrogenase (Cabrini *et al.*, 1969, 1970), various dehydrogenases (Butcher, 1972; Nolte & Pette, 1972), glucose-6-phosphatase (Abrahamson, Rigatuso & Lazarow, 1969) and α-glycerophosphate dehydrogenase (Hess & Pearse, 1961).

A biomedical application of microdensitometry which may prove to be of considerable practical importance is the group of methods introduced by Chayen and his colleagues (Chayen, Bitensky & Daly, 1974) for the assay of various hormones, including ACTH, LH, TSH and LATS. Suitable 'target' cells known to react to the particular hormone in question are treated *in vitro* with the experimental serum, and stimulation is assessed by an appropriate quantitative cytochemical test and microdensitometry. The methods are extemely sensitive, since only a few target cells and correspondingly minute quantities of test serum are required. Each system is calibrated by the use of control sera containing known amounts of hormone, and the specificity and stoichiometry of the cytochemistry itself are accordingly less crucial than in most other applications of microdensitometry.

Most objects which are suitable for integrating microdensitometry absorb a proportion of the light passing through them. If the absorption is effectively complete the specimen appears black, i.e. with an absorbance of infinity, and the integrated absorbance of a microscopical field containing even one such particle would in theory be infinite. It therefore appears at first sight that highly absorbing or opaque particles could not usefully be studied by integrating microdensitometry. The technique might also seem not to be applicable to metallic specimens, which appear dark in transmitted illumination because of reflection, not absorption, and to which the Bouguer–Beer law does not apply. In practice, however, integrating microdensitometry can sometimes yield useful information with both metallic and effectively opaque particles such as result from certain histochemical enzyme procedures. Consider a microscopical field containing a number of small opaque particles, such as autoradiographic silver grains. In the presence of uncompensated glare the apparent absorbance of an opaque object is finite, e.g. with 5 % glare the apparent absorbance cannot exceed 1.3. Even if glare is negligible, or electronically compensated, the diameter of the measuring spot used in a scanning microdensitometer is usually greater than that of a silver grain in an autoradiograph; during a scan the measuring spot therefore never lies over a grain without overlapping onto the empty background, and the apparent transmission will at no time be zero. The absorbance measurable by a microdensitometer is also in practice restricted to a maximum (e.g. 1.3 in the Vickers M85). Due to these instrumental factors, the apparent integrated absorbance of even a relatively large opaque particle will be finite, and the apparent integrated absorbance of a microscopical field containing many such particles will be approximately proportional to the number present. The number of silver grains per unit area of an autoradiograph is approximately proportional to the exposure to which the emulsion was subjected, provided this is kept low. Integrating microdensitometry is therefore a useful method for quantitating moderately exposed autoradiographs, and it can be shown empirically that the method is even applicable to autoradiographs too heavily exposed for valid measurement with incident-light reflectance techniques (Goldstein & Williams, 1971).

Advantages, disadvantages and limitations of integrating microdensitometry

In contrast with in-vitro spectrophotometry, microdensitometry permits the retention of morphological structure and spatial relationships in the material studied. The substance measured can therefore be attributed to particular tissues, cells or intracellular organelles, and differences between different cell types or regions in an apparently homogeneous organ can readily be recognised. Another major advantage of microdensitometry is its sensitivity. Given sufficiently high local concentrations, precise microdensitometric measurements may be made of picogram quantities of light-absorbing material, e.g. the haemoglobin content of a single red blood cell (*c.* 30×10^{-12} g) can be measured without difficulty. Conventional spectrophotometry is, however, superior for the measurement of relatively large volumes of dilute solutions, and colour-generating analytical reactions in solution are in general better understood, more reliable and available for a greater variety of substances than equivalent tests modified for cytochemical use with insoluble substrates.

Microdensitometry and conventional, visual microscopy should be regarded not as rival but as complementary techniques. If the material studied is either inherently coloured, or has been stained by a technique of known specificity, microdensitometry enables objective comparisons to be made between different tissues or cells, and in favourable cases microdensitometric assessment of absolute quantities of light-absorbing material may be possible. Visual, qualitative microscopy is however indispensable, and it should be remembered that although the human eye is unreliable for the assessment of the concentration or total amount of stain present in a microscopical object, under some conditions it is quite good at *matching* intensities. Thus a comparison microscope is a useful adjunct to an integrating microdensitometer, enabling rapid and easy comparisons to be made e.g. of differently stained but morphologically similar structures on separate microscope slides. If no microdensitometer is available, it is sometimes possible to use a comparison microscope to obtain a surprising amount of quantitative data (see e.g. Goldstein, 1963, 1965).

Regarding the limitations of integrating microdensitometry, both random and systematic instrumental errors are largely avoidable with modern equipment. Care should, however, be taken to define the purpose of the investigation, since to minimise time and effort the precision and accuracy of the data obtained should not greatly exceed what is required. Thus relatively few and simple measurements suffice to allocate a Feulgen-stained nucleus into one or other ploidy class, but more refined techniques (including e.g. compensation for residual distribution error and glare) may be necessary to assess the significance of minor deviations of apparent Feulgen-DNA contents from expected modal values.

For maximum instrumental precision and accuracy the specimen absorbance should be moderate, preferably about 0.4 or 0.5. Even if the absorbance does not exceed the measuring range of the instrument (1.0 to 1.5 in most cases) serious systematic errors can occur with highly absorbing specimens, e.g. from uncorrected or even overcorrected glare. The reproducibility of repeated measurements is also low both with very dark and very pale objects. Undesirably high absorbances can usually be decreased by using a wavelength of light some distance from the peak of the specimen's absorption spectrum, but this may cause difficulties if very pale objects must also be measured. Under these circumstances measurements may have to be made at two different wavelengths; this procedure may also be adopted to measure specimens doubly stained with dyes having different absorption spectra, but being time-consuming and of relatively low precision should be avoided if possible.

Problems associated with specimen preparation are in general far less tractable than merely instrumental difficulties in microdensitometry. Before measurement cells and tissues are usually fixed or frozen, smeared, squashed or sectioned, and subjected to a variety of other procedures in addition to staining. Uncontrolled loss of the material of interest may occur at any stage in the proceedings, and relatively subtle changes such as partial denaturation of proteins or alterations in the activities of enzymes may significantly affect the results. If sections are used, technical problems may arise due to the thickness of the specimen affecting the focus, and difficulties of interpretation may result from cells or organelles being sliced through and hence incompletely present on a given slide. On the other hand if smears or squashes are employed the spatial relationships between cells are inevitably disturbed, and it is not always easy to identify particular cell types in a smear from a complex organ. In addition there may be a selective loss, e.g. of particularly large or fragile cell types, during the smearing or preliminary isolation procedures.

As already pointed out, one of the main advantages of microdensitometry over other quantitative methods is the retention of spatial relationships in the material studied. Apparent localisations can, however, be misleading. The intracellular polarisation of glycogen in conventionally fixed and alcohol-dehydrated cells is a familiar histochemical artifact, and the apparent localisation of alkaline phosphatase in or on cell nuclei, suggested by early histochemical work, is now recognised to result from diffusion of the enzyme and/or its reaction product.

Apart from possible problems of incomplete preservation and inaccurate localisation, it might be thought easy to adapt an established histochemical staining method for quantitative microdensitometric use. This is unfortunately not true, and although it is statistically simple to show that a given method is precise (i.e. reproducible), painstaking and laborious investigation is necessary to prove its quantitative validity (Table 2). The distinctions between instrumental 'precision' and 'accuracy', and between stain 'selectivity' and 'specificity' have already been discussed, but even when the accuracy of the machine and the specificity of the histochemical procedure have been established, experimental conditions must be devised which ensure that the dye uptake is proportional to the amount of substrate over the range of concentrations likely to be encountered. This step in the validation of a quantitative method is often extremely difficult. Model systems consisting of known concentrations of chemically defined substances bound to or dispersed in films of known composition have been used successfully

Table 2. *Fallacies in quantitation*

Precision	=	Accuracy of measurement
Accuracy	=	Selectivity of staining
Selectivity	=	Specificity
Specificity	=	Stoichiometry
Stoichiometry	=	Constant absorptivity

(e.g. van Dalen, Ahsmann & Van Duijn, 1970; Tas, Oud & James, 1974), but it is not always clear that the physico-chemical conditions present in a fixed cell or tissue are adequately simulated either by solutions or by insoluble but relatively homogeneous films. If stoichiometry of the dye uptake can be established, it is still necessary to know the absorptivity (proportionality constant in the Bouguer–Beer law) if absolute amounts of a coloured material are to be deduced from absorbance data. The absorptivity of a dye attached to a fixed cell is not necessarily identical with that of the same substance in solution, and a further complication arises from the fact that with many dyes absorptivity is a function of concentration (cf. metachromasia). The question of absorptivity also arises in connection with the measurement of inherently coloured materials such as haemoglobin; the concentration of haemoglobin inside a red cell is many times that customarily used for in-vitro measurements, and extrapolations from one set of conditions to the other may not be valid.

At the present time very few histochemical techniques have been investigated sufficiently thoroughly for their behaviour, vagaries and limitations to be understood with any confidence, and the quantitative validation of staining methods currently presents the major obstacle to a more widespread application of integrating microdensitometry.

Future developments in integrating microdensitometry

Commercially available microdensitometric apparatus has already reached a considerable degree of sophistication, and it does not appear that fundamental improvements are imminent. Instrumental developments in the near future may, however, be expected in two, perhaps mutually incompatible, directions. On the one hand, relatively inexpensive microdensitometers for routine use may have increased automation, speed and ease of operation. Automatic setting of background absorbance levels, for example, would speed up many types of work considerably and also diminish operator error. On the other hand, with increasing understanding of and confidence in microdensitometry, there may be a greater demand for versatile and powerful instruments, with which the various residual microdensitometric errors can be corrected and measurements of improved precision and accuracy made. Such instruments may also be capable of a variety of additional measuring tasks, such as reflectance, ultraviolet and fluorescence work, image analysis and integrating microinterferometry.

From the point of view of applications, we have already seen that the most important difficulty in the way of integrating microdensitometry is the lack of reliable staining methods. The establishment of the validity of additional techniques is therefore the most urgent task facing quantitative microscopists. As a corollary to this we must accept that some staining procedures at present widely accepted as valid may not stand the test of time, and have to be discarded for quantitative work.

The ultimate success of microdensitometry will not perhaps be measurable by the number of available microdensitometers, or even by the number of workers specialising in the field. On the contrary, microdensitometry should gradually cease to be the prerogative of the specialist, and become a routine laboratory tool of acknowledged usefulness in a wide variety of biological investigations.

References

Abrahamson, D. E., Rigatuso, J. L. & Lazarow, A. (1969). Cytophotometric quantitation of glucose-6-phosphatase activity in rat liver. *Journal of Histochemistry and Cytochemistry*, **17**, 107–9.

Altman, F. P. (1975). Quantitation in histochemistry: a review of some commercially available microdensitometers. *Histochemical Journal*, **7**, 375–95.

Altman, F. P., Moore, D. S. & Chayen, J. (1975). The direct measurement of Cytochrome P450 in unfixed tissue sections. *Histochemistry*, **41**, 227–32.

Bácsy, E. & Rappay, Gy. (1971). Microspectrophotometric data on the kinetics of acid phosphatase activity in cells removed from the peritoneal cavity of the rat. *Histochemical Journal*, **3**, 193–9.

Bedi, K. S. & Goldstein, D. J. (1974). Cytophotometric factors causing apparent differences between Feulgen DNA contents of different leukocyte types. *Nature, London*, **251**, 439–40.

Böhm, N., Sprenger, E., Schluter, G. & Sandritter, W. (1968). Proportionalitätsfehler bei der Feulgen-Hydrolyse. *Histochemie*, **15**, 194–203.

Butcher, R. G. (1972). Precise cytochemical measurement of neotetrazolium formazan by scanning and integrating microdensitometry. *Histochemie*, **32**, 171–90.

Cabrini, R. L., Frasch, A. C. C. & Itoiz, M. E. (1975). A quantitative microspectrophotometric study of the lead precipitation reaction for the histochemical demonstration of acid phosphatase. *Histochemical Journal*, **7**, 419–26.

Cabrini, R. L., Itoiz, M. E., Mayo, J., Smolko, E. E. & Klein-Szanto, A. J. P. (1970). Microspectrophotometric study of histoenzymic reactions in rat epidermis subjected to 0.5 Mrads of deuteron radiation. *International Journal of Radiation Biology*, **18**, 415–21.

Cabrini, R. L., Vinuales, E. J. & Itoiz, M. E. (1969). A microspectrophotometric method for histochemical quantitation of succinic dehydrogenase. *Acta histochemica*, **34**, 287–91.

Chayen, J., Bitensky, L. & Daly, J. R. (1974). Cytochemical bioassay of hormones. *Life Sciences*, **15**, 911–201.

van Dalen, J. P. R., Ahsmann, W. B. A. M. & van Duijn, P. (1970). A method for the determination of the molar extinction coefficient of structure-linked chromophores. *Histochemical Journal*, **2**, 329–42.

Friede, R. L. & Knoller, M. (1965). Quantitative tests of histochemical methods for phosphomonoesterases in brain tissue. *Journal of Histochemistry and Cytochemistry*, **13**, 125–32.

Goldstein, D. J. (1963). An approach to the thermodynamics of histological dyeing, illustrated by experiments with azure A. *Quarterly Journal of Microscopical Science*, **104**, 413–39.

Goldstein, D. J. (1965). A further note on the measurement of the affinity of a dye (Azure A) for histological substrates. *Quarterly Journal of Microscopical Science*. **106**, 299–306.

Goldstein, D. J. (1970). Aspects of scanning microdensitometry. I. Stray light (glare). *Journal of Microscopy*, **92**, 1–16.

Goldstein, D. J. (1971). Aspects of scanning microdensitometry. II. Spot size, focus and resolution. *Journal of Microscopy*, **93**, 14–42.

Goldstein, D. J. (1975). Aspects of scanning microdensitometry. III. The monochromator system. *Journal of Microscopy*, **105**, 33–56.

Goldstein, D. J. & Williams, M. A. (1971). Quantitative autoradiography: an evaluation of visual grain counting reflectance microscopy, gross absorbance measurements and flying-spot microdensitometry. *Journal of Microscopy*, **94**, 215–39.

Hess, R. & Pearse, A. G. E. (1961). Histochemical and homogenisation studies of mitochondrial α-glycerophosphate dehydrogenase in the nervous system. *Nature, London*, **191**, 718–19.

James, V. & Goldstein, D. J. (1974). Haemoglobin content of individual erythrocytes in normal and abnormal blood. *British Journal of Haematology*, **28**, 89–101.

Jongkind, J. F., Reuser, A. J. J. & Galjaard, H. Quantitative enzyme assays on single cells. *Proceedings of the Royal Microscopical Society*, **10**, 280.

Keilin, D. (1925). On cytochrome, a respiratory pigment, common to animals, yeast, and higher plants. *Proceedings of the Royal Society of London, Series B*, **98**, 312–39.

Liebman, P. A. & Entine, G. (1964). Sensitive low-light-level microspectrophotometer: detection of photosensitive pigments of retinal cones. *Journal of the Optical Society of America*, **54**, 1451–9.

MacMunn, C. A. (1886). Researches on myohaematin and the histochaematins. *Philosophical Transactions of the Royal Society of London*, Series B, **177**, 267–98.

Mayall, B. H. & Mendelsohn, M. L. (1970*a*). Deoxyribonucleic acid cytophotometry of stained human leukocytes. II. The mechanical scanner of CYDAC, the theory of scanning photometry and the magnitude of residual errors. *Journal of Histochemistry and Cytochemistry*, **18**, 383–406.

Mayall, B. H. & Mendelsohn, M. L. (1970*b*). Errors in absorption cytophotometry: some theoretical and practical considerations. In *Introduction to Quantitative Cytochemistry*, ed. G. L. Wied & G. F. Bahr, vol. 2, pp. 171–97. New York & London, Academic Press.
Mendelsohn, M. L. (1966). Absorption cytophotometry: comparative methodology for heterogeneous objects, and the two-wavelength method. In *Introduction to Quantitative Cytochemistry*, ed. G. L. Wied, pp. 202–14. Academic Press, New York & London.
Nolte, J. & Pette, D. (1972). Microphotometric determination of enzyme activity in single cells in cryostat sections. I. Application of the gel film technique to microphotometry and studies on the intralobular distribution of succinate dehydrogenase and lactate dehydrogenase activities in rat liver. *Journal of Histochemistry and Cytochemistry*, **20**, 567–76.
Pearse, A. G. E. (1968). *Histochemistry*, 3rd edn, vol. 1. J. & A. Churchill, London.
Pearse, A. G. E. (1972). *Histochemistry*, 3rd edn, vol. 2. Churchill-Livingstone, Edinburgh & London.
Ruch, F. (1972). Quantitative Histochemie der Proteine. *Acta histochemica, Supplement*, **12**, 57–60.
Tas, J., Oud, P. & James, J. (1974). The Naphthol Yellow S stain for proteins tested in a model system of polyacrylamide films and evaluated for practical use in histochemistry. *Histochemistry*, **40**, 231–40.
Thorell, B. (1958). Cell studies with microspectrography. In *Advances in Biological and Medical Physics*, ed. C. A. Tobias & J. H. Lawrence, vol. 6, pp. 95–119. Academic Press, New York & London.
Walker, P. M. B. (1956). Ultraviolet absorption techniques. In *Physical Techniques in Biological Research*, vol. 3, ed. G. Oster & A. W. Pollister, pp. 402–87. Academic Press, New York & London.
Welford, W. T. (1972). On the relationship between the modes of image formation in scanning microscopy and conventional microscopy. *Journal of Microscopy*, **96**, 105–7.
Wied, G. L. (ed.) (1966). *Introduction to Quantitative Cytochemistry*. Academic Press, New York & London.
Wied, G. L. & Bahr, G. F. (eds.) (1970). *Introduction to Quantitative Cytochemistry*, vol. 2. Academic Press, New York & London.

D. J. GOLDSTEIN

Scanning microinterferometry

General principles

Most phenomena in microscopy can be explained, or at least described, by assuming light to consist of transverse waves with the properties of direction, amplitude, frequency or wavelength*, phase and polarisation (Fig. 1). The visible intensity of light is proportional to the square of its amplitude, and wavelength is perceived as colour. Neither polarisation nor phase are seen as such by the unaided human eye, but the phase relationship of two beams of light can be rendered visible by *interference*: the instantaneous amplitudes of two waves add algebraically (i.e. taking account of positive and negative signs), and the amplitude and corresponding intensity of the resulting light depend on the amplitudes and relative phase of the interfering beams (Fig. 2).

For interference between two beams to be demonstrable their relationship must remain unaltered over a relatively long period. Light waves, however, occur in discrete packets called photons, and there is no consistent relationship between the phase, plane of polarisation and to some extent the wavelength of photons emitted from different regions of a heated filament, or of photons coming successively from a given region. Beams of so-called *incoherent* light from two conventional tungsten lamps therefore cannot interfere, and their combined intensity is simply the sum of the original intensities. Beams of mutually *coherent* light, capable of interference, can be obtained by splitting a single beam (Fig. 3). Under these circumstances light from a given point in the source reaches its destination along two different routes, and the relative phase of the beams remains unaltered by random changes occurring simultaneously in both.

Consider the interference of two equally intense beams of coherent light, originally derived from a single light source (Fig. 3). At a given point in space

* For light of a given colour the frequency (number of vibrations per unit time) is constant, but the wavelength and velocity are both inversely proportional to the refractive index of the medium. When the medium is unspecified, 'wavelength' implies the wavelength in air or vacuum.

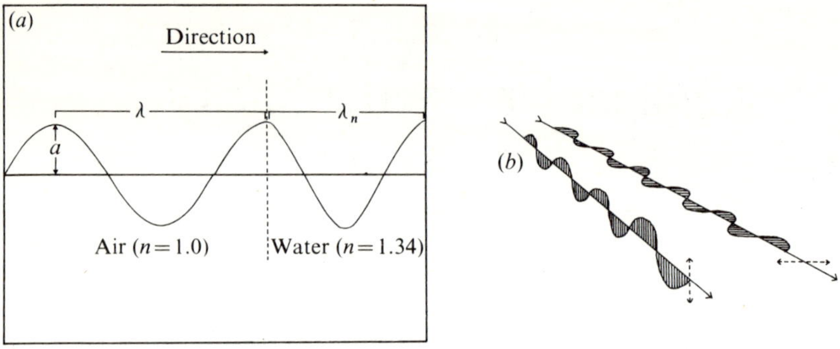

Fig. 1. (*a*) Wavelength (λ) and maximum amplitude (*a*) of light. The wavelength depends on the refractive index (*n*) of the medium. At a given moment the amplitude may be positive (as illustrated), zero or negative. (*b*) Polarisation refers to the plane of vibration of the transverse wave when looked at from the direction in which the light is travelling.

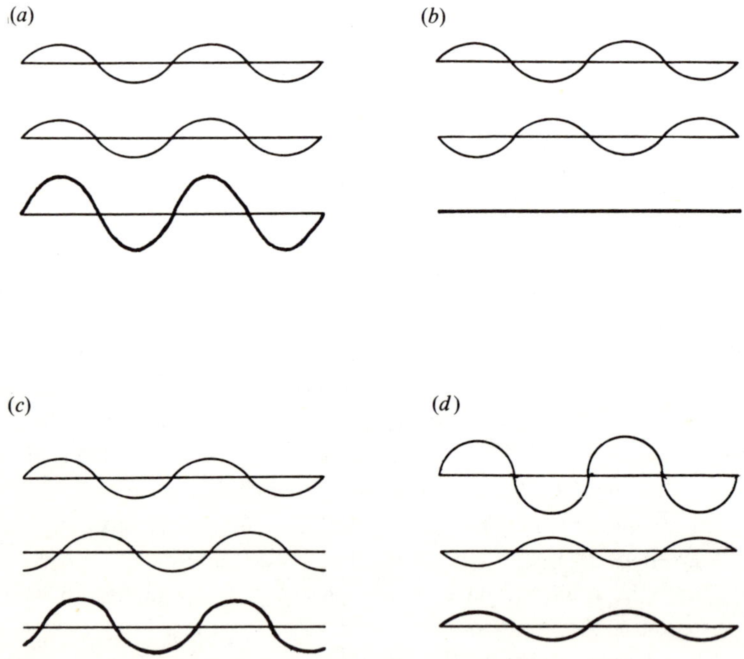

Fig. 2. Interference of coherent light waves (thin lines) to give a resultant (thick line). (*a*) Waves of equal amplitude, exactly in phase, give a resultant with the same phase and twice the amplitude. (*b*) Waves of equal amplitude, exactly half a wavelength out of phase, give resultant with zero amplitude (i.e. total destructive interference). (*c*) Waves of equal amplitude, λ/4 out of phase, give a resultant of intermediate phase with an amplitude $2^{0.5}$ times that of either of the interfering waves. (*d*) Waves of unequal amplitude, exactly out of phase, give a finite resultant.

the relative phase of the re-combined beams depends on the length of time each has been travelling. This is proportional to the *optical path*, defined as the product of the geometrical distance and the refractive index of the medium. If the two beams have travelled identical total optical paths they arrive exactly in phase, and the amplitude and intensity of the resulting light are respectively twice and four (i.e. 2^2) times that of either beam alone (Fig. 2*a*). This is *constructive* interference. One beam may, however, have been relatively retarded, e.g. by passing through a region of high refractive index (Fig. 3*b*). In the special case where the difference between the optical paths (the *optical path difference*, or OPD) is half a wavelength, the beams arrive exactly out of phase with each other and the resulting amplitude and intensity are zero (Fig. 2*b*). Such *destructive* interference can be total only if the interacting beams are of equal amplitude (cf. Figs. 2*b* and 2*d*). With phase differences other than zero and half a wavelength the results of interference are intermediate between the extreme cases just described (Fig. 2*c*).

Interference causes a re-distribution but no net loss or gain of light energy, and constructive interference at one point is invariably accompanied by destructive interference elsewhere. The complementary nature of constructive and destructive interference is sometimes evident as alternating bright and dark interference fringes, but in other instances the bright and dark regions may not readily be visible simultaneously (Fig. 3).

Interference techniques enable the detection and measurement (Fig. 3*c*) of minute differences in such characteristics of macroscopic specimens as thickness and refractive index. An interference system of the Mach–Zehnder type, similar in principle to that illustrated in Fig. 3, can be incorporated in a specially constructed microscope having two objectives and two condensers, one pair to visualise the specimen and the other to transmit the reference beam of coherent light. Such an instrument is manufactured (e.g. by Leitz), and it is possible with it both to visualise transparent microscopic objects and to measure a number of important cell and tissue properties (for reviews of microinterferometry see Davies, 1958; Hale, 1958; Barer, 1959; Ross, 1967). Most commercial interference microscopes, however, employ a polarising interference system of the Jamin–Lebedeff type, which permits the relatively inexpensive construction of compact, mechanically stable and precise instruments with only a single condenser and objective.

In a typical polarising interference microscope (Fig. 4) linearly polarised light is split by a calcite plate in the microscope condenser into two coherent beams, with planes of polarisation at right-angles to each other. One beam is deviated laterally, and from the calcite beam splitter the beams emerge parallel and separated by a distance of some tens or hundreds of micro-

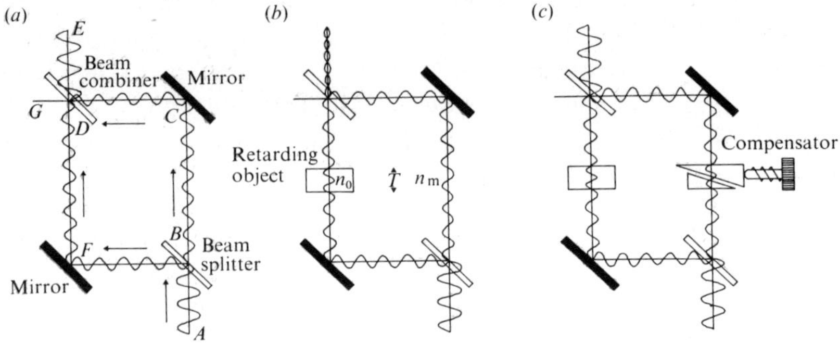

Fig. 3. Functioning of an interferometer of the Mach–Zehnder type. Light from a source *A* is split into coherent beams at *B*, and after total reflection at mirrors *C* and *F* respectively the beams are reunited at the semi-reflecting mirror *D*. (*a*) The optical paths travelled by the beams are equal, and constructive interference occurs at *E*. (*b*) Retardation of one beam by an object of thickness *T* and refractive index n_0 has introduced a phase difference between the beams of $(Tn_o - Tn_m) = \lambda/2$, where n_m is the refractive index of the medium traversed by the other beam. Destructive interference occurs at *E*. (*c*) Equality of the pathlengths has been restored by a variable compensator. In this diagram the law of the conservation of energy superficially appears to be violated. In fact, energy 'missing' from point *E* in diagram (*b*) reappears as constructive interference at point *G*, while in diagrams (*a*) and (*c*) destructive interference occurs at *G* due to a half-wavelength phase difference between the beams passing respectively along paths *ABCDG* and *ABFDG*. Without going into the explanation for this phase difference in detail, it should be noted that a half-wavelength phase change occurs whenever light in a given medium is reflected from the surface of a medium of higher refractive index; a single such change is undergone by beam *ABCDG*, but two by beam *ABFDG*.

metres.* A birefringent (half-wave) plate rotates the plane of polarisation of each beam through 90°, and the signal (or 'specimen') and reference beams now pass respectively through the microscopic object and through a nearby region of empty mounting medium. The beams enter a second calcite plate incorporated in the microscope objective; because of the reversal of the planes of polarisation caused by the half-wave plate, the beam which was previously deviated laterally is now unaffected, and vice versa. The beams are thus re-combined, but are still polarised at right-angles to each other and may in addition be out of phase due to relative retardation or advancement of the

* In the 'differential' interference microscope designed by Nomarski and marketed by various manufacturers the beam separation is only a fraction of a micrometre. This system is, however, in general useful only for qualitative observations.

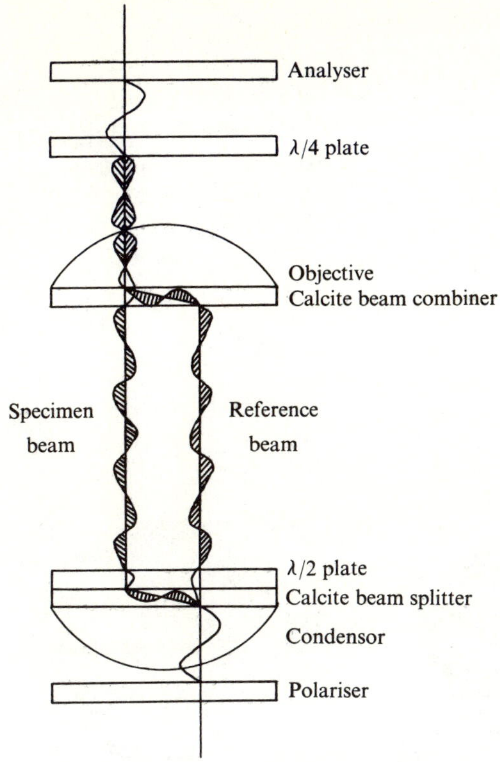

Fig. 4. The Jamin–Lebedeff polarising interference system as used in microscopy. A Sénarmont quarter-wave plate analysing system is shown, but other types of compensator can also be used. The light source and eye-piece (not shown) are as in a conventional microscope.

specimen beam by the object. Various devices can be employed to bring the beams into a common plane of polarisation and enable interference to take place. In the Sénarmont system (Fig. 4) a suitably oriented quarter-wave-retarding birefringent plate converts the two plane-polarised beams into a single beam, the plane of polarisation of which is rotated through an angle directly proportional to the phase difference between the interfering beams (for a detailed explanation see Hallimond, 1970). The light then passes through an analyser which can be rotated so as either to pass or to reject light of any chosen plane of polarisation. Thus a dark specimen may be seen on a bright background or vice versa (Figs. 5 and 6), and the OPD of a specimen can be measured by noting the angular rotation of the analyser necessary to convert an image with a maximally dark background to one where the image of the specimen itself is black.

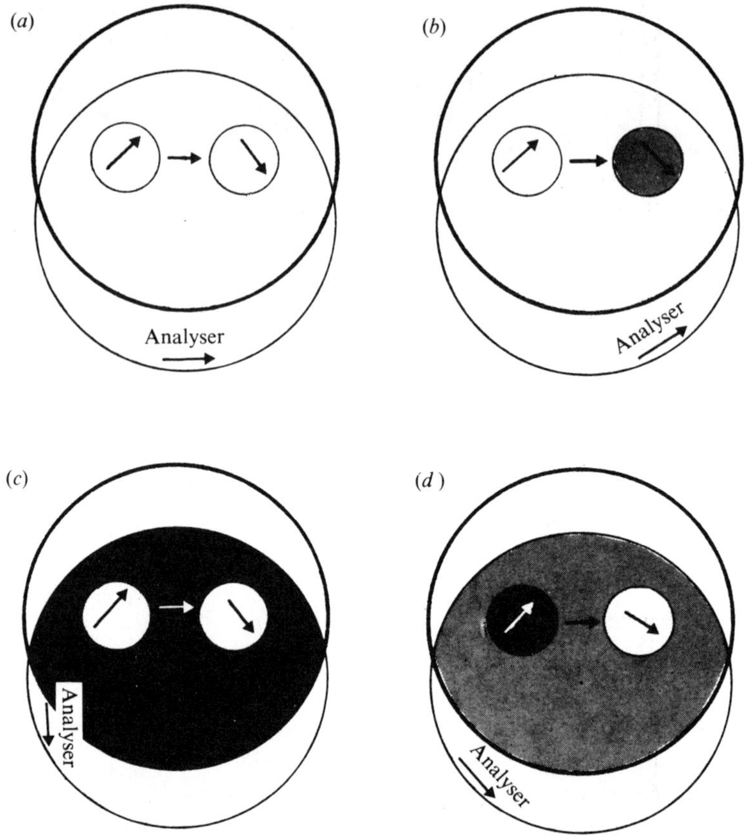

Fig. 5. Appearance of the image plane of a polarising interference microscope, as seen with various settings of the analyser. The planes of polarisation of the background, image and 'ghost' image of the specimen are indicated by arrows; the transmission plane of the analyser is similarly shown. Note that the rotation of the planes of polarisation of the image and ghost image are equal but opposite. (*a*) Background bright, both images somewhat darker. (*b*) Image of specimen bright, background darker, ghost image darker still. (*c*) Background completely extinguished; image and ghost image relatively bright. (*d*) Image completely extinguished; background brighter, and ghost image brighter still.

A drawback of the Sénarmont analyser system is its limited measuring range. With optical path differences of more than one wavelength only the fractional part of a wavelength is measured; for example, actual OPDs of 0.4, 1.4 and 2.4 wavelengths are all apparently 0.4. With a specimen of high OPD the integral number of wavelengths must therefore be ascertained by other means, e.g. a fringe eyepiece. With this device a birefringent quartz wedge is superimposed on the image of the microscopical field, and a series of dark fringes is seen joining points of equal OPD. Thus a sort of contour map is obtained, and the OPD of a specimen can be estimated approximately by the displacement of the fringes crossing it (Fig. 6*c*).

Due to the beam-splitting (or beam-combining) calcite plate in the objective, all parts of the microscopical field are seen doubled in the image plane of a polarising interference microscope. The visible image of the specimen is in fact formed by the interference of two superimposed images, one of the transparent specimen and the other the displaced image of a reference area of empty medium. The displaced image of the specimen itself may also be seen to one side of the 'true' image, superimposed either on empty medium or the image of another object. In certain makes of interference microscope (e.g. Vickers, Zeiss), the displaced image is astigmatic and blurred, and hence relatively unobtrusive – it is commonly referred to as a 'ghost'.

The plane of polarisation of light in the ghost image is rotated through an angle equal but opposite in sign to that of the true image (Fig. 5). With certain settings of the analyser the true and ghost images therefore appear with opposite contrast. The ghost image is itself of only minor significance, but it is important that the ghost image of some other specimen should not fall on the true image of the specimen being studied. If this were to happen it would mean that the reference beam is passing not through empty medium but through the other object, and interpretation of the interference image may be impossible. With an object having a diameter greater than the separation of the specimen and reference beams, the true and ghost images of the specimen will necessarily overlap to some extent, and only the non-overlapping part of the image will be amenable to quantitative study. This is the major inherent drawback of the Jamin–Lebedeff system.

It may be noted that ghost images are not a result of the law of the conservation of energy; light 'missing' from a dark image in a polarising interference microscope does not re-appear as a bright ghost, but is dissipated as heat in the microscope analyser. Ghost images are in this respect different from the familiar 'halos' of phase contrast microscopy, which immediately surround the image of the specimen and are an inevitable consequence of the re-distribution of energy involved in the process of image formation. Phase contrast halos can considerably impair the clarity of the image seen, and the absence of halos in interference microscopy is a substantial advantage of the system for qualitative work. In addition, spatially extended objects are more faithfully represented by interference microscopy, phase contrast tending to overemphasise the edges of the images of homogeneous specimens. Interference microscopes are also in general more easily adjusted to give optimum images of specimens of widely differing retardation. Since the first commercial

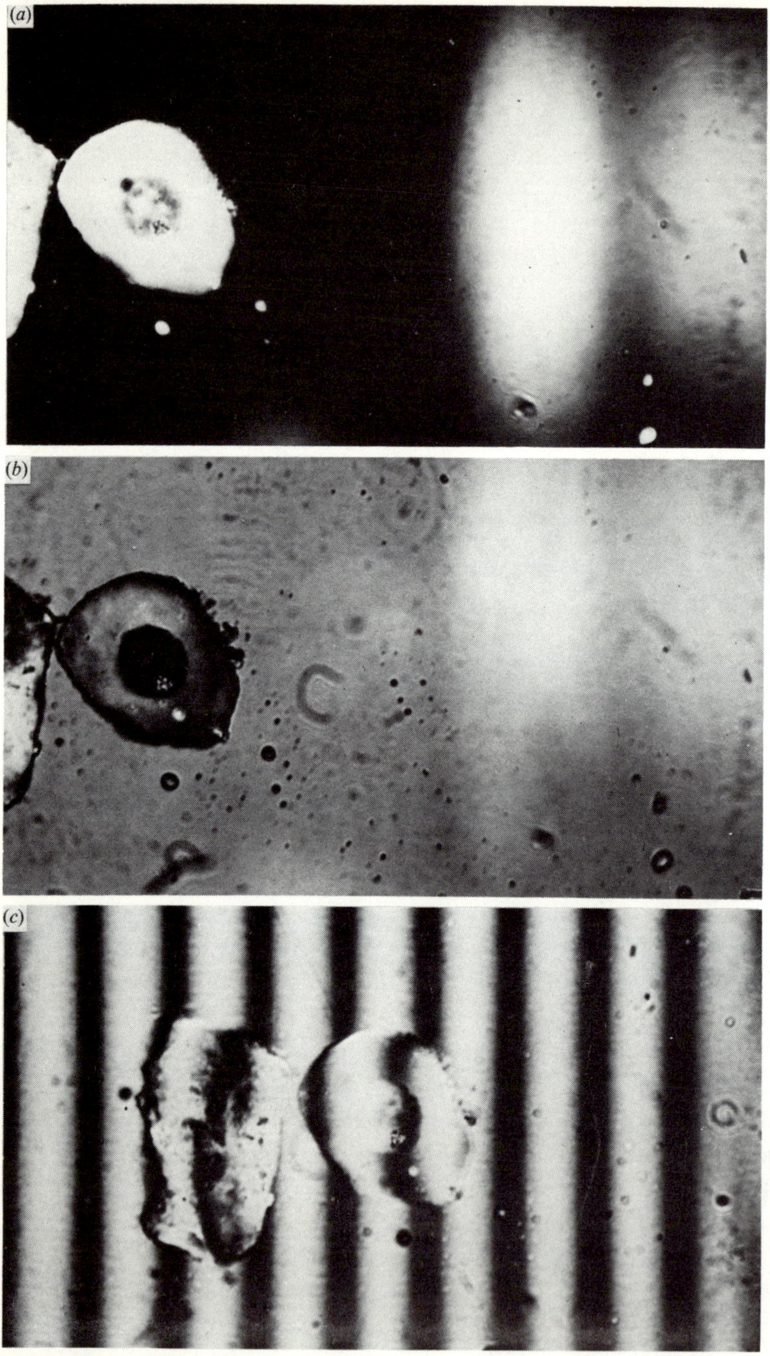
(a)
(b)
(c)

introduction of practical, high-power interference microscopes about 25 years ago, interference microscopy has accordingly been successfully used for qualitative studies on both living and fixed cells and tissues.

The most important applications of interference microscopy are, however, quantitative, and some discussion of the significance of OPD measurements is necessary. If T is the thickness of the specimen, and n_o and n_m are respectively the refractive indices of the object and the surrounding medium (Fig. 3),

$$\begin{aligned} \text{OPD} &= Tn_o - Tn_m \\ &= T(n_o - n_m). \end{aligned} \tag{1}$$

The value of n_m being known or readily measurable (e.g. by an Abbe refractometer), n_o can be calculated from an OPD measurement provided T is known, and vice versa. It is not possible to evaluate both T and n_o from a single OPD measurement, but if measurements are made in two different media of refractive indices n_{m1} and n_{m2} respectively, we have

$$\text{OPD}_1 = T(n_o - n_{m1}) \tag{2}$$

and

$$\text{OPD}_2 = T(n_o - n_{m2}). \tag{3}$$

Both T and n_o can be derived from the above simultaneous equations, but this 'two medium' method is more often applicable to inorganic specimens or fixed tissues than to living cells, which usually require to be immersed in an isotonic aqueous medium.

The possibility of calculating the dry mass of a living cell from a single OPD measurement was pointed out independently by Barer & Joseph (1954) and Davies, Wilkins, Chayen & La Cour (1954). The refractive index of an aqueous solution is linearly related to the concentration of solute, the proportionality factor α ('specific refractive increment') being customarily defined as the increase in refractive index caused by the addition of 1 g of solute to 100 ml of solution. The specific refractive increment is nearly the same for all the major constituents of living cells, being about 0.0018. The difference $(n_o - n_m)$ between the refractive indices of a cell and of water is therefore approximately proportional to the mass of solute per unit volume of

---

Fig. 6. Isolated pig epidermal cells mounted in water, viewed by polarising interference system. (*a*) Analyser set to give black background. Bright images of two cells are seen on the left, and the corresponding bright ghost images, blurred and out of focus, on the right. (*b*) Analyser set to give minimum intensity of the nucleus of one of the cells. (*c*) Appearance with fringe eyepiece. The retardation of the cell nucleus is slightly less than that of the cytoplasm, which is just under half a wavelength.

the cell, and since thickness (cm) multiplied by mass per unit volume (g cm^{-3}) is equivalent to mass per unit area (g cm^{-2}), the OPD of a cell mounted in an aqueous medium is proportional to the dry mass of the cell per unit projected area. If the OPD and the projected area A of a homogeneous cell are measured respectively in μm and μm^2, and if mass is expressed in pg (i.e. 10^{-12} g), then

$$\text{Dry mass per unit area} = \text{OPD}/100\alpha \tag{4}$$

and

$$\text{Total dry mass} = A\cdot\text{OPD}/100\alpha. \tag{5}$$

The factor 100 is necessary because of the differences in the units used for area, mass and OPD on the one hand and α on the other.

Despite these interesting possibilities of quantitative interference microscopy, practical applications have until recently been severely handicapped by the fact that most biological specimens are notably non-uniform. The OPD can be measured with reasonable accuracy at any given point in a specimen, but in order to obtain a reliable estimate of the total dry mass of a heterogeneous specimen the mean OPD must be established by measurements at many different points. This procedure is too slow and tedious to be practical with conventional interference microscopes. Clearly a device is required which can rapidly and objectively measure the product of the projected area and the mean OPD (i.e. the 'integrated OPD', or IOPD) of a microscopic specimen. Several such instruments have been built and described, but of these only the Vickers M86 scanning and integrating microinterferometer appears to be commercially available at the present time. The remainder of this chapter will accordingly be devoted to the M86.

Description of Vickers M86 polarising microinterferometer

The instrument (Figs 7–9) is a polarising microinterferometer of the Jamin–Lebedeff type, equipped with double-refracting optics in principle identical with those of the visual interference microscope designed by F. H. Smith and originally marketed by Messrs Baker (now incorporated in Vickers Instruments Ltd). The M86 has, however, facilities for the automatic, electronic measurement of OPD and for the automatic scanning and integrating measurement of IOPD. The apparatus moreover possesses all the microdensitometric capabilities of the Vickers M85 scanning and integrating microdensitometer (see previous chapter). Descriptions of the M86 can be found e.g. in Smith (1967, 1972) and Goldstein & Hartmann-Goldstein (1974), and the present discussion covers only points essential for an understanding of the capabilities and limitations of the instrument.

Fig. 7. The Vickers M86 scanning and integrating microinterferometer. The laser light source for interferometric measurements, and the stabilised tungsten lamp for microdensitometry are housed in the horizontal unit at the top of the microscope column. An automatic camera is shown attached to the instrument.

When in visual use, light from a conventional tungsten lamp below the microscope condenser passes up through a Jamin–Lebedeff system consisting of a polariser, interference condenser, matched objective, quarter-wave plate and analyser to reach the eyepiece and the human eye. The lateral separation of the beams in the plane of the object is about 80, 160 and 330 μm respectively for the $\times 75$ (water-immersion), $\times 40$ and $\times 10$ objectives. The functioning of the system in this mode of operation has already been described. When the instrument is used in the automatic measuring mode the light path is reversed. Light from a helium–neon laser (wavelength 632.8 nm) at the top of the instrument passes through a phase-modulating unit (see below) to reach the objective, where it is split into specimen and reference beams. The specimen beam is focused by the objective into a measuring spot about 0.4 μm in diameter (when using the $\times 75$ objective); this spot can be moved in the specimen plane by the automatic or manually controlled movement of a pair of mirrors above the objective. The reference and specimen

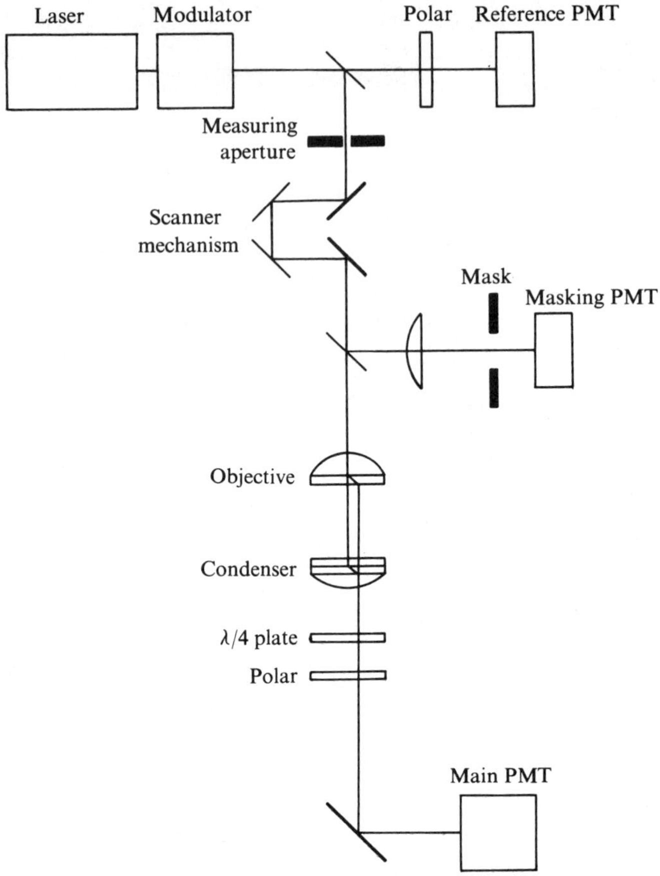

Fig. 8. Optical system of the Vickers M86 microinterferometer (highly simplified). The scanner mechanism consists of two fixed and two vibrating mirrors. The measuring aperture, the mask of the indirect specimen-defining system and the specimen itself are in optically conjugate planes (see text). When used for visual observation, light comes from a conventional tungsten lamp (not shown) below the condenser; a prism above the objective deflects the light into a binocular eyepiece at the side of the microscope column. PMT, photomultiplier tube.

beams are collected and united by the condenser, pass through a quarter-wave plate and a polarising filter and are eventually received by a photomultiplier tube. The electronic phase-measuring device operates almost as if it were a conventional Sénarmont system having an analyser rotating at 5000 rev s^{-1}. This effect is actually achieved by the phase-modulating unit, which contains a 1600-line circular diffraction grating which is mechanically revolved at 3.125 rev s^{-1}. The intensity of the modulated light reaching the photomultiplier tube fluctuates at 10 kHz, and the phase of the similarly fluctuating electrical output of the tube is a function of the retardation of the specimen beam relative to the reference beam in the plane of the specimen. The phase timing of the main photomultiplier tube output is compared electronically with the output of a reference photocell which samples the light leaving the modulator unit, and the final result is that the OPD of the specimen relative to the background is measured 10000 times per second.

With a stationary measuring spot, automatic averaging of the 10000 OPD measurements over a period of precisely 1 s gives a highly reproducible measurement of the OPD. The IOPD of a discrete, heterogeneous object can be measured in approximately 3 s, by electronic integration of the instantaneous OPD measurements made during a two-dimensional scan of the specimen by the moving measuring spot. Although the total area scanned each time is rectangular, precise delimitation of the specimen area of interest is made possible by an indirect 'masking' system; the only OPD measurements integrated are those made while the measuring spot lies within an area defined by a mechanical mask of variable size and shape (Fig. 9).

Measuring range

As with a conventional Sénarmont system the maximum OPD measurable with the M86 is theoretically one wavelength. Electronic and other noise in the system in practice limits the effective range of the instrument to about 0.9 wavelengths, and care must be taken to avoid specimens having optical path differences greater than this. Preliminary inspection with a fringe eyepiece is therefore advisable. In addition, use may be made of an area-measuring device incorporated in the M86 which automatically records the area of specimen having an OPD greater than a pre-set threshold value. A specimen with an OPD greater than unity can generally be detected by setting this threshold to (say) 0.95 wavelengths, since during a scan of such an object the instantaneous OPD will usually at some point lie between 0.95 and 1.0 wavelengths, and a finite area reading will therefore be obtained. If specimens with an OPD greater than unity are encountered it may be possible to reduce the OPD to a measurable value by substituting a mounting medium of higher refractive index. For example, anhydrous glycerol with a refractive index of about 1.47 might be used in place of water (refractive index about 1.34). Alternatively, it is sometimes feasible to reduce the OPD of a specimen by gentle compression, or by the use of thin sections.

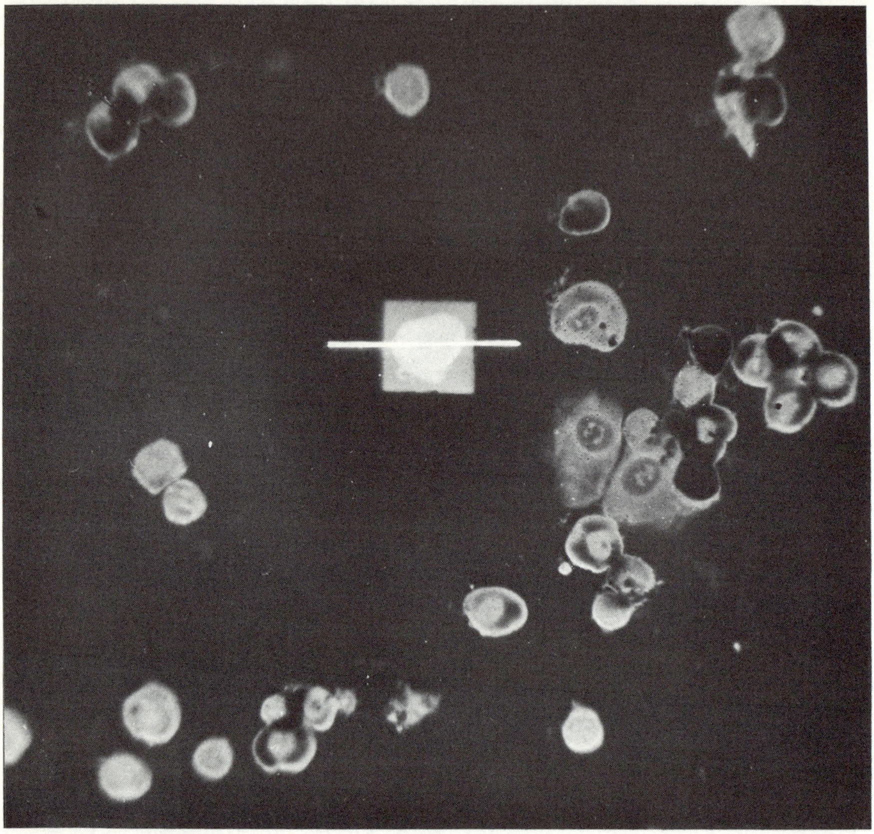

Fig. 9. Photomicrograph of pig epidermal cells in water, visualised with Vickers M86 microinterferometer. The cell to be measured is here delimited by a rectangular mask (other shapes can also be used). The transverse line is a single raster line of the moving measuring spot; during a complete scan a fast (x) movement combined with a slow (y) movement of the spot covers a two-dimensional area. Only measurements made within the mask are integrated.

Reproducibility (precision)

With its measuring spot stationary, the Vickers M86 can make repeated OPD measurements reproducible to less than 0.001 wavelengths. This is roughly an order of magnitude better than is achievable with a conventional, visual interference microscope employing a Jamin–Lebedeff system and a Sénarmont compensator. To illustrate the precision of the M86 with a concrete example, Araldite has a refractive index of about 1.54. A 0.04 μm ultra-thin Araldite section for electron microscopy has therefore an OPD in air of about 0.04 (1.54–1.0)/0.632 = 0.034 wavelengths of helium–neon light, and the thickness of such a section can be measured with a coefficient of variation of about ±3%. Repeated measurements at a given site will of course improve the reproducibility still further.

The precision of scanning and integrating measurements of IOPD depends on the area and retardation of the specimen. Coefficients of variation of about 5 % and 10 % respectively are found with repeated measurements of the IOPD of human red cells (Fig. 10*a*) and mouse sperm heads (Fig. 10*b*) mounted in water, and better results are obtainable with larger or more retarding objects. When the M86 is used as a microdensitometer the coefficient of variation of repeated integrated absorbance measurements of a Feulgen-stained mouse sperm head is about 1 %, so that the instrument appears somewhat less precise in the interference mode. The standard error of a mean value is, however, inversely proportional to the square root of the number of observations, so that the IOPD of an object can be measured with any precision desired. Thus the IOPD of an erythrocyte can in 2 min be measured at least eight or nine times, together with the same number of 'blank' measurements of empty background, and the dry mass can be estimated with a precision of better than ±2 %.

Accuracy

By simply increasing the number of observations one can improve the precision but not the accuracy (i.e. freedom from systematic error) of a measurement. There is no easy method of detecting, let alone eliminating systematic instrumental errors, but three possible sources of error in scanning and integrating microinterferometry have been investigated.

Under some conditions an almost sinusoidal deviation from instrumental linearity is seen with the M86, the error in the apparent OPD depending on the actual value measured (Goldstein & Hartmann-Goldstein, 1974). Under the worst conditions a maximum error of about ±0.02 wavelengths is found with specimens having a true OPD of around 0.25 or 0.75 wavelengths, but the error is negligible with objects with an OPD near zero or 0.5 wavelengths. The maximum error can be considerably reduced by careful adjustment of the optical system and by limitation of the effective numerical aperture of the condenser. It is probable that the error arises from factors connected with the birefringent calcite beam splitters (see an explanation by F. H. Smith, quoted by Goldstein & Hartmann-Goldstein, 1974), and that an analogous error occurs with all interference microscopes employing the Jamin–Lebedeff

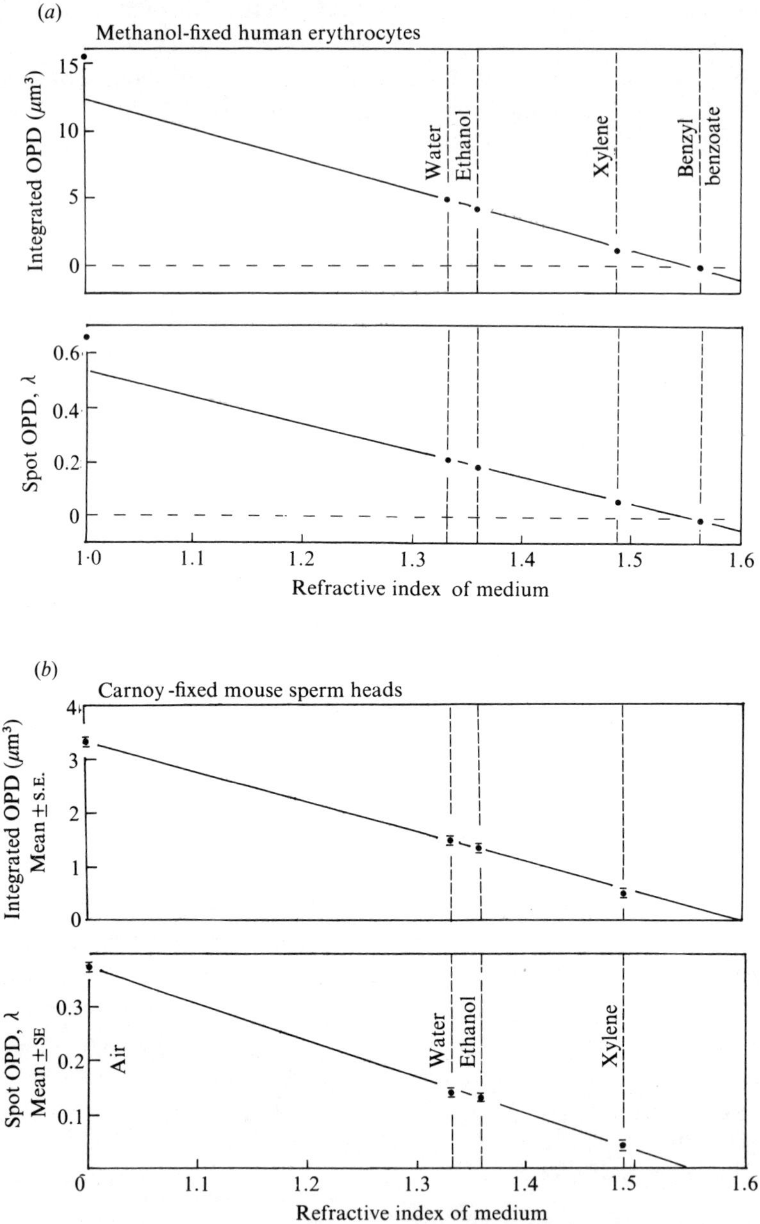

Fig. 10 (*a*) and (*b*). OPD (stationary measuring spot) and IOPD (scanning and integrating) measurements of cells mounted in various media. According to elementary theory both OPD and IOPD should be linearly related to the refractive index of the mounting medium. In the case of a mouse sperm head the OPD in water was slightly but significantly less than expected on the basis of the other measurements.

system; the deviation from linearity was perhaps first noticed with the Vickers M86 because of the relatively high precision of that instrument.

In marked contrast with scanning and integrating microdensitometry, IOPD measurements are extremely insensitive to the size and precise focus of the measuring spot in the specimen plane (Goldstein & Hartmann-Goldstein, 1974). This fortunate circumstance makes it possible to measure the IOPD and hence dry mass even of specimens considerably thicker than the depth of focus of the objective used (provided their OPD does not exceed one wavelength).

Glare (stray light) in the optical system might be expected to affect OPD and IOPD measurements in much the same way as in microdensitometry (see previous chapter). It therefore appears prudent to reduce glare as much as possible by using suitably located field stops to restrict the specimen area illuminated. In practice, however, glare appears to have little or no effect on interferometric measurements with the M86. This is perhaps because stray light reaching the photomultiplier tube has been largely depolarised and demodulated by multiple reflections, and hence does not contribute significantly to the signal detected by the phase-sensitive electronic system.

Applications

The Vickers M86 has been available for too short a time for much work to have been carried out with the instrument, and it is not yet possible to evaluate with any confidence the place of the instrument in cell biology. A certain amount of research has, however, been carried out using non-commercial integrating microinterferometers; thus Petersson (1968) studied the dry mass of pancreatic islet cells with an instrument originally developed by Caspersson and his colleagues, and a few other articles based on work with the same or similar instruments have been published by workers in Stockholm and elsewhere. In addition, the Vickers M86 has been applied to a variety of problems. These include the relationship of dry mass to the cell dimensions of macrophages (Mayhew & Williams, 1973), and changes in dry mass occurring under various conditions in tumour cells (LePoint, Bassleer & de Paermentier, 1974; LePoint, Bassleer, de Paermentier &

This was probably due to lateral swelling of the specimen, the effect not being seen with IOPD measurements of the whole cell. Both OPD and IOPD measurements of a human red cell in air were higher than expected; this was apparently due to uptake of water by the hygroscopic specimen, as the measurements rose to the observed values within a few minutes of removal of the slide from a drying oven.

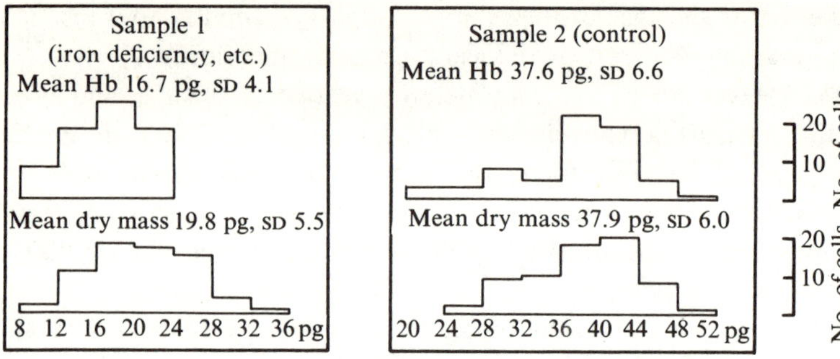

Fig. 11. Comparison of microinterferometric dry mass and microdensitometric haemoglobin measurements of erythrocytes in abnormal (Sample 1) and normal (Sample 2) human blood. In blood from a case of iron deficiency anaemia both the corpuscular haemoglobin-content and dry mass were reduced. Direct comparison of the dry mass and haemoglobin measurements is not possible, as the absolute figures depend on the values assumed for the absorptivity of haemoglobin and the specific refractive increment of the red cell cytoplasm.

Lhoest-Gauthier, 1974), nerve cells (Noel-Courtey & LePoint, 1974), fibroblasts (de Paermentier, LePoint & Bassleer, 1974) and human epidermal cells (Yardley & Goldstein, 1976). More recent work, mostly unpublished at present, includes studies on the dry mass of *Drosophila* polytene nuclei isolated (Hartmann-Goldstein, 1972) from animals reared under different environmental conditions, changes in cartilage matrix following surgical trauma (Matthews & Goldstein, 1976), differences in the dry mass of individual human erythrocytes (Fig. 11), and changes in the dry mass of preimplantation mouse embryos. In all these investigations the dry mass data obtained could not readily have been procured by any other technique.

Although it seems probable that the major use of integrating microinterferometry in biology will be in the measurement of cell and tissue dry mass, other possible applications exist. Thus microinterferometry with a conventional, non-scanning instrument has been used to quantitate cell and tissue phosphatase activity (Barter, Danielli & Davies, 1955; Casselman, 1959); the mass of phosphate precipitated during incubation of smears or sections in a Gomori-type medium was measured directly, without the need to convert the reaction product into an opaque form. The elimination of several steps from the normal histochemical procedure reduces possible sources of error, and the technique has the advantage over absorbance methods of not damaging the specimen, and thus allowing repeated measurements to be made at a given site. The generally heterogeneous character of the precipitate is a handicap for work with a non-scanning instrument, but would be of little consequence with the Vickers M86.

By making microinterferometric dry mass estimations both before and after selective enzymic or chemical extraction procedures, attempts have been made to measure the cellular or nuclear content of such substances as protein, lipid and nucleic acid (Davies, Deeley & Denby, 1957; Bibbiani, Tongiani & Viola-Magni, 1969). Although there is often some doubt regarding the completeness and/or specificity of the removal, the ease and precision of scanning microinterferometry may well cause a revival of interest in this type of method, which is valuable as a completely independent check on the microdensitometric estimation of specifically stained material.

When used in the non-scanning mode of operation, point measurements of OPD can be made with the Vickers M86 as with other interference microscopes. The precision, speed and objectivity of the M86, however, make certain applications much easier, and perhaps even feasible for the first time. Thus knowledge of the precise thickness of microtome sections is often highly desirable in such work as quantitative autoradiography, visible-light microdensitometry and quantitative electron microscopy; as already pointed out it is possible with the M86 to measure even ultrathin sections with precision. Other examples of point OPD measurements with the M86 are shown in Fig. 10(*a*) and (*b*).

Advantages, disadvantages and prospects of integrating microinterferometry

Microinterferometric measurements with the Vickers M86 are rapid, objective, require no special skill, and have the great advantage over most other quantitative methods in cell biology of being completely non-destructive. Repeated measurements can therefore be made even on living cells, and the same specimens subsequently used for other types of investigation. Small changes in total dry mass or in mass per unit projected area are readily detected, and to obtain the dry mass of a specimen in absolute units it is only necessary to assume a value for the mean specific refractive increment. The dry mass of tissues, cells and cell organelles is not only of interest in itself, but can provide a base-line for the standardisation of other quantitative estimations.

The disadvantages and limitations of microinterferometry with the M86 are partly technical and partly more fundamental. The necessity for a specimen-free reference area near the object may make it impossible to measure closely packed smears or large microscopic sections, unless parts of the preparation are removed mechanically. The same factor may make it difficult to measure the thickness of some parts of an ultrathin section mounted on a conventional electron microscope grid, since the relevant reference areas may be occupied by an opaque grid bar (this is manifested by the ghost image of the bar overlying the distinct image of the specimen). With experience such inconveniences can, however, usually be overcome.

The fact that the measuring range of the M86 is limited to one wavelength usually presents no difficulty in the measurement of specimens such as blood cells, oral squames and cultured fibroblasts mounted in an aqueous medium. Thicker objects, such as mammalian ova, may however have to be either compressed or mounted in a medium of relatively high refractive index. The use of non-aqueous media unfortunately raises the problem that in certain cases the value of the specific refractive increment is somewhat controversial. Very varied figures are for example given in the literature for estimating the dry mass of cells measured in anhydrous glycerol. It seems that more work is required in this field, and in the meantime it may sometimes be best to report the results of measurements in non-aqueous media in relative rather than absolute units.

The very general applicability of dry mass estimations is paradoxically probably the most basic limitation of scanning and integrating microinterferometry. Because dry mass measurements are inherently non-specific, an increase in one cell component may be masked by a decrease in another, and net changes may be impossible to interpret. It is conceivable that microinterferometry may prove most useful in cell biology when used in conjunction with less general but more specific techniques such as microdensitometry, autoradiography and microfluorometry. These radically different approaches should be regarded as being complementary rather than competitive with each other.

There now appear to be no significant technical obstacles to the interferometric measurement of cellular dry mass and mass per unit area. The next few years should see a much more general availability of the requisite instrumentation, and with wider use of integrating microinterferometry it should be easier to assess the importance of the technique and the permanence of its place in cell biology.

References

Barer, R. (1959). Phase interference and polarizing microscopy. In *Analytical Cytology*, 2nd edn, ed. R. C. Mellors, pp. 169–272. McGraw-Hill, New York.

Barer, R. & Joseph, S. (1954). Refractometry of living cells. I. Basic principles *Quarterly Journal of Microscopical Science*, **95**, 399–423.

Barter, R., Danielli, J. F. & Davies, H. G. (1955). A quantitative cytochemical method for estimating alkaline phosphatase activity. *Proceedings of the Royal Society of London*, Series B, **144**, 412–26.

Bibbiani, C., Tongiani, R. & Viola-Magni, M. P. (1969). Quantitative determination of the amount of DNA per nucleus by interference microscopy. *Journal of Cell Biology*, **42**, 444–51.

Casselman, W. G. B. (1959). Some applications of interference microscopy in enzyme histochemistry, with particular reference to alkaline phosphatase. *Experimental Cell Research, Supplement*, **7**, 28–39.

Davies, H. G. (1958). The determination of mass and concentration by microscope interferometry. *General Cytochemical Methods*, **1**, 55–161.

Davies, H. G., Deeley, E. M. & Denby, E. F. (1957). Attempts at measurement of lipid, nucleic acid and protein content of cell nuclei by microscope-interferometry. *Experimental Cell Research, Supplement*, **4**, 136–49.

Davies, H. G., Wilkins, M. H. F., Chayen, J. & La Cour, K. F. (1954). The use of interference microscope to determine dry mass in living cells as a quantitative cytochemical method. *Quarterly Journal of Microscopical Science*, **95**, 271–304.

de Paermentier, F., LePoint, A. & Bassleer, R. (1974). Mesure par micro-interferometrie à balayage de la masse sèche totale de cellules normales ou cancereuses cultivées *in vitro*. *Bulletin de l'Association des Anatomistes*, **58**.

Goldstein, D. J. & Hartmann-Goldstein, I. J. (1974). Accuracy and precision of a scanning and integrating microinterferometer. *Journal of Microscopy*, **102**, 143–64.

Hale, A. J. (1958). *Interference Microscope in Biological Research*. E. & S. Livingstone Ltd. Edinburgh,

Hallimond, A. F. (1970). *The Polarizing Microscope*, 3rd edn. Vickers Ltd, York.

Hartmann-Goldstein, I. J. (1972). Simple method for isolating nuclei from individual salivary glands of *Drosophila*. *Drosophila Information Service*, **49**, 129–30.

LePoint, A., Bassleer, R. & de Paermentier, F. (1974). Etude cytochimique quantitative des proteines totales dans les cellules tumorales d'Ehrlich. *Comptes rendus de l'Académie des Sciences, Paris, Series D*, **279**, 1229–31.

LePoint, A., Bassleer, R., de Paermentier, F. & Lhoest-Gauthier, M. P. (1974). Dosages cytophotometriques et microinterferometriques des proteines totales de cellules tumorales d'Ehrlich cultivées *in vitro*. *Bulletin de l'Association des Anatomistes*, **58**.

Matthews, J. G. & Goldstein, D. J. (1976). Remodelling in immature hyaline cartilage: changes in matrix dry mass. *Journal of Anatomy*, in press.

Mayhew, T. & Williams, M. A. (1973). The mass and size of normal and activated macrophages – studies with a scanning interferometer. *Experientia*, **29**, 80–1.

Noel-Courtey, B. & LePoint, A. (1974). Masse sèche et teneur proteines totales de cellules nerveuses de moelle spinale cultivées *in vitro*. *Bulletin de l'Association des Anatomistes*, **58**.

Petersson, B. (1968). The dry mass of the pancreatic B-cells in relation to their content of secretion granules. *Histochemical Journal*, **1**, 55–8.

Ross, K. F. A. (1967). *Phase Contrast and Interference Microscopy for Biologists*. Edward Arnold, London.

Smith, F. H. (1967). A photo-electronic phase-measuring microscope. *Journal of the Royal Microscopical Society*, **87**, 165–9.

Smith, F. H. (1972). A laser illuminated scanning microinterferometer for determining the dry mass of living cells. *Microscope*, **20**, 153–60.

Yardley, H. J. & Goldstein, D. J. (1976). Changes in dry weight of human epidermal cells undergoing keratinization as determined by scanning interference microscopy. *British Journal of Dermatology*, in press.

H. Y. ELDER

The potential of the scanning transmission electron microscope in biology

For many years now the scanning electron microscope (SEM) has been a familiar and valued tool for the biologist, giving up to 40× better resolution and a greater depth of focus (up to several thousand times) at high magnification than the light microscope. By its use a large amount of valuable information on surface fine structural detail has been accumulated. Since in a scanning microscope the image is built up point-by-point in a time sequence as a finely focused beam is scanned raster-fashion over the object, resolution is critically dependent upon the minimum size of the focused beam. In practical terms the probe current must be greater than 10^{-11} A and with a conventional heated tungsten source this limits resolution to about 5 nm. Since conventional transmission electron microscopes (TEM) can achieve resolutions an order of magnitude greater than SEMs, there have until recently been few attempts to construct a scanning transmission electron microscope (STEM). Although it has been known for some time that field emission cathodes should be capable of producing an electron source some 10^4–10^6 times brighter than thermionic cathodes, early attempts, some twenty years ago (Gabor, 1974), to make a field emission gun (FEG) were unsuccessful because the ultrahigh vacuum (10^{-8}–10^{-10} torr) necessary for a FEG could not be maintained. Following the advent of the ion pump the field emission gun was developed in the 1960s by Albert Crewe in Chicago and the STEM came of age spectacularly with the first acknowledged pictures of single atoms (Crewe, Wall & Langmore, 1970; Crewe & Wall, 1970). Currently commercial STEM instruments with resolution better than 0.5 nm are available. Summarised below are some of the advantages for biologists which the STEM is thought to offer, with particular reference to quantitative and analytical microscopy. Although there are at least three commercially available STEM instruments at the present time the high cost of these (£100000–£200000) inevitably means, as Ferrier (this volume) points out, that evaluation will be slow. Figs. 1 and 2 show a photograph of the type of instrument discussed by Burge (this volume) and a diagram of its optical column.

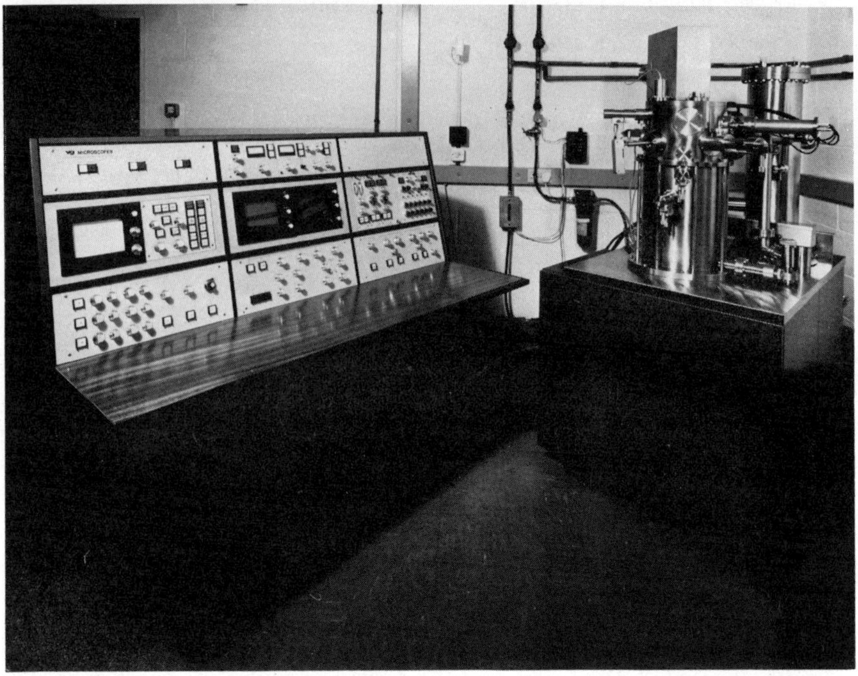

Fig. 1. Photograph of an HB5 STEM instrument similar to that used by Burge (this volume), courtesy of VG Microscopes. The control and display console is to the left and the microscope column and vacuum pumps to the right. The field emission gun is in the housing at the base of the column and the electron spectrometer is contained in the rectangular chamber at the top of the column.

Advantages of STEM with FEG

Images from thick sections

A fundamental defect of the TEM instrument is chromatic aberration which limits resolution because the transmitted electrons must be imaged by lenses with high chromatic aberration coefficients. The major source of the fault lies in the fact that a large proportion of the incident electrons (especially in histological material of low atomic number) interacts in various ways with the atoms and molecules of the specimen, losing some energy in the process. Such inelastically scattered electrons, as well as exhibiting a spectrum of energy loss, are of longer wavelength than the incident electrons and are focused closer to the specimen than higher energy electrons. We thus have an electron optical analogue of chromatic aberration in the light microscope and it is mainly specimen induced. The resolution loss is particularly prominent in thick specimens and this is the major factor in limiting most TEM applications to ultrathin sections. Because in STEM the monoenergetic electrons are focused before interacting with the specimen and imaging lenses are therefore not necessary, imaging of thick specimens is much improved. It has been calculated that for a given accelerating voltage STEM may be able to produce comparable resolution to TEM from a specimen some 3–10 times thicker (Crewe & Groves, 1974; Groves, 1975). Further references to this potentially important development in biological microscopy are given by Burge (this volume).

Instability effects on the image

The STEM image is built up point-by-point in a time sequence and because of the intensity of the probe and the sensitivity of the detectors (the STEM scintillation detectors are some 10^3 times more sensitive than the TEM fluorescent screen) the time necessary to record an individual image element is often less than 10^{-4} s. This means that mechanical and electrical instabilities, which are a serious cause of resolution limitation in TEM, cause only distortion in STEM.

Image brightness and rotation

Since magnification is determined by the ratio of scanned area of specimen to display screen size and the electron intensity of the imaging cathode ray tube is independent of the specimen-illuminating electron beam of the microscope, magnifications from a few hundred times, theoretically to infinity, with rotation-free images of undiminished brightness can be obtained.

Reduced contamination

Scanned biological specimens in TEMs with scanning attachments and SEMs (in which the vacuum is of the order 10^{-5}–10^{-6} torr) are subject to rapid contamination, partly from the instrument itself (vacuum oils and seals, etc.) and partly from the specimen. The great disadvantage of this is illustrated in the dramatic pictures of contamination build-up shown by Hutchinson (this volume) in situations where, for microanalysis, it is necessary to keep a static probe on areas for times up to several minutes. The ultra-high vacuum technology necessitated by the FEG ensures examination in a 'clean' vacuum system and the contamination of specimen origin can be greatly reduced by a variety of methods, e.g. baking the specimens for an hour or more at 50 °C, 'fixing' migration-prone contaminants by electron beam or ultraviolet irradiation, or, for radiation-sensitive or heat-labile specimens, examination on a cold stage (*c.* −120 °C). See comments in the chapters by Appleton, Burge and Hutchinson (this volume) for further details.

Microanalytical modes

One of the greatest potential benefits of the FEG–STEM system lies in the area of microanalysis. First, there is the potential of ionisation loss electron spectroscopy which is the subject of the chapter by Ferrier (this volume). Secondly, in the field of X-ray microanalysis, dealt with in the chapters by Chandler and Hutchinson (this volume), FEG–STEM offers further advantages. Having no imaging lenses, the area in the vicinity of the specimen is much less congested than in a TEM (see Fig. 2) and space is available to optimise specimen-detector geometry to achieve maximum sensitivity and signal-to-noise ratio. Also an intense probe of only a few nanometres diameter (< 2–3 nm) allows the possibility of detecting X-ray signals from areas perhaps smaller than 5 nm in ultrathin sections.

Thirdly, the small intense probe will allow electron diffraction to be performed on diameters down to 5 nm or less. Though the FEG permits small, intense probe currents for diffraction imaging (some 50 nA), the current is spread over all the image points and part of the advantage is lost by having to record the image point-by-point in a time sequence by means of the post-specimen microdiffraction scan coils (Fig. 2). Of various solutions considered perhaps the simplest is the suggestion to insert a fluorescent screen and camera for direct imaging and recording of diffraction patterns in the STEM. Since facilities for electron energy loss spectroscopy, X-ray microanalysis and electron diffraction can all be accommodated easily, the FEG-STEM promises to be a microanalytical instrument of unrivalled power for biologists in the future.

Contrast and STEM image modes

For the most common requirements of resolution and magnification in biology, namely shape discrimination of stained cellular organelles and membrane system, the most important mechanism of contrast formation is amplitude contrast, which arises by the differential removal of electrons, giving rise, therefore, to dark areas in the image. These are produced by thick and dense specimen regions and, very importantly for the biologist studying ultrathin sections, by heavy atoms ('atomic number contrast') introduced as selectively distributed 'stain' within the lower atomic number matrix. The

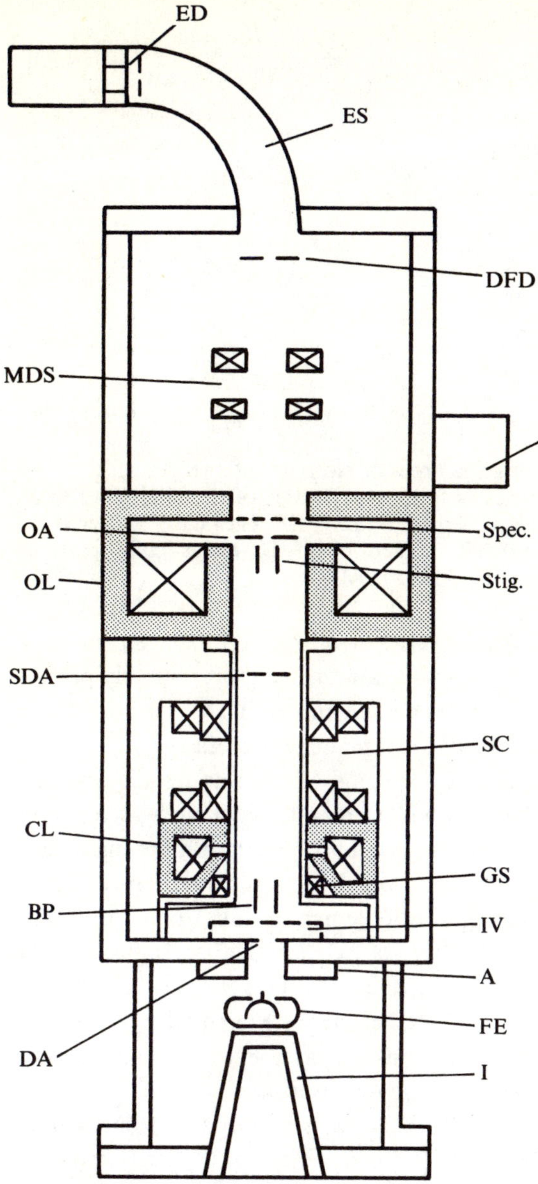

Fig. 2. Schematic illustration of the column of the HB5 STEM, courtesy of VG Microscopes. Abbreviations as follows: A, anode; BP, beam blanking plate; CL, condenser lens; DA, differential aperture; DFD, detector aperture and dark-field detector; ED, exit slit and detector; ES, electron spectrometer; FE, field emission point and extraction electrode; GS, gun stigmator; I, 100 kV insulator; IV, column isolation valve; MDS, microdiffraction scan coil assembly; OA, objective aperture; OL, objective lens; SA, specimen airlock; SC, scan and alignment coil assembly; SDA, selected area diffraction aperture; Spec., specimen; Stig., stigmator.

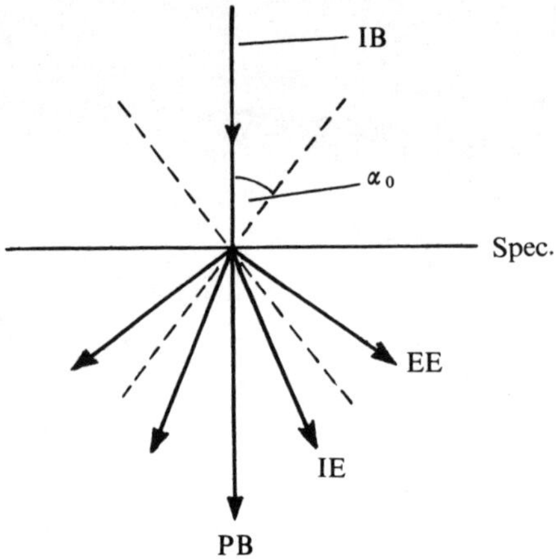

Fig. 3. Diagram illustrating the types of electron interaction with the specimen. Abbreviations as follows: α_0, angle of the illuminating cone formed by the objective lens in STEM; EE, elastically scattered electrons (diffracted beam); IB, incident beam; IE, inelastically scattered electrons; PB, primary beam; Spec., specimen.

scattering capability of atoms increases with atomic number (Z), or more strictly, the total elastic scattering cross-section, σ_e, is approximately proportional to $Z^{\frac{4}{3}}$. Amplitude contrast is increased in the TEM by introducing smaller apertures in the objective back focal plane which cut out more of the diffracted beams (with loss, of course, of some of the fine detail carried in the widely diffracted beams) (Fig. 3). Phase contrast, which arises from some retardation of part of the beam so that it is out of phase with the undeviated primary beam, becomes progressively more important in high-resolution studies and is critically dependent upon the degree of underfocus. In STEM the three image components can be separately detected (Fig. 4). An annular detector collects elastically scattered electrons (N_e). This is essentially a dark-field image (DF). The inelastically scattered electrons, which have lost energy, can be separated from the primary electrons by passing these beams through an electron spectrometer (see the chapter by Ferrier, this volume).

The 'no-energy-loss' signal (N_o), which contains mainly the primary beam is the STEM equivalent of bright-field (BF). The inelastically scattered electrons (N_i) can be used to form an energy loss image. Micrographs can be formed from each or all channels and often most usefully from specific combinations, e.g. the ratio N_e/N_i gives a signal intensity approximately proportional to Z (Crewe *et al.*, 1970; Langmore & Wall, 1970; Crewe, 1974).

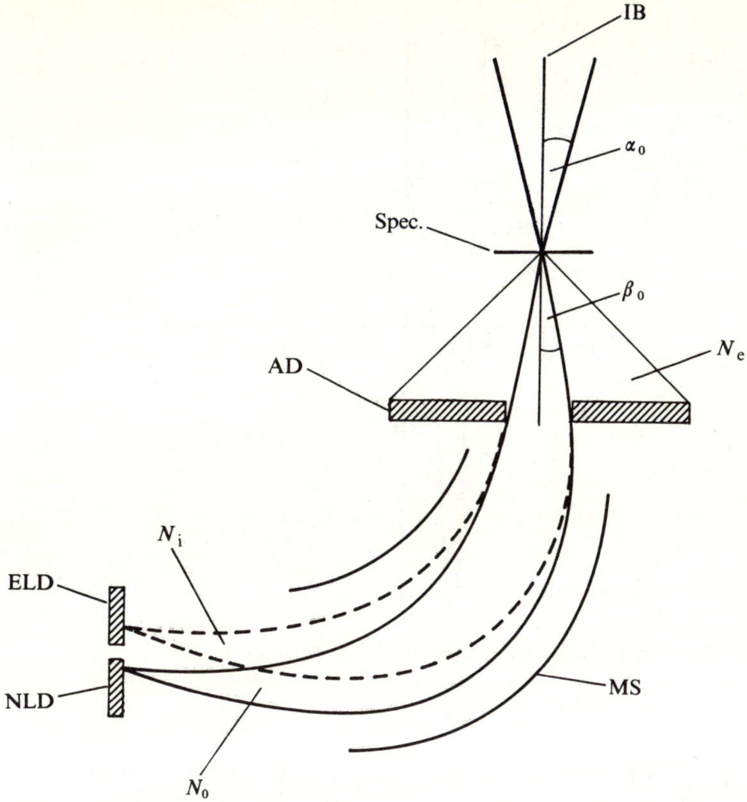

Fig. 4. Diagram illustrating in STEM the separate methods of detecting the elastically scattered electrons (N_e), zero loss electrons (N_o) and inelastically scattered electrons (N_i). Abbreviations as follows: α_0, probe angle of the objective lens; AD, annular detector; β_0, the acceptance angle subtended by the spectrometer; ELD, energy loss detector; IB, incident beam; MS, magnetic spectrometer; NLD, zero loss detector; Spec., specimen.

Bright-field amplitude contrast in the STEM

With respect to the often-cited principle of reciprocity (Burge, this volume, Fig. 3) governing the illumination and image-detecting systems of TEM and STEM instruments, Burge shows that, while in principle the two systems are similar, the different values for illumination and detection angles which must apply in practice give the STEM instrument a much less favourable signal-to-noise ratio than TEM in the bright-field imaging mode. If the objective aperture angle, α_0, is comparable to the angle β_0, which is the illumination angle in TEM or the detector angle in STEM, then amplitude contrast predominates over phase contrast; this is the situation for bright-field STEM imaging at

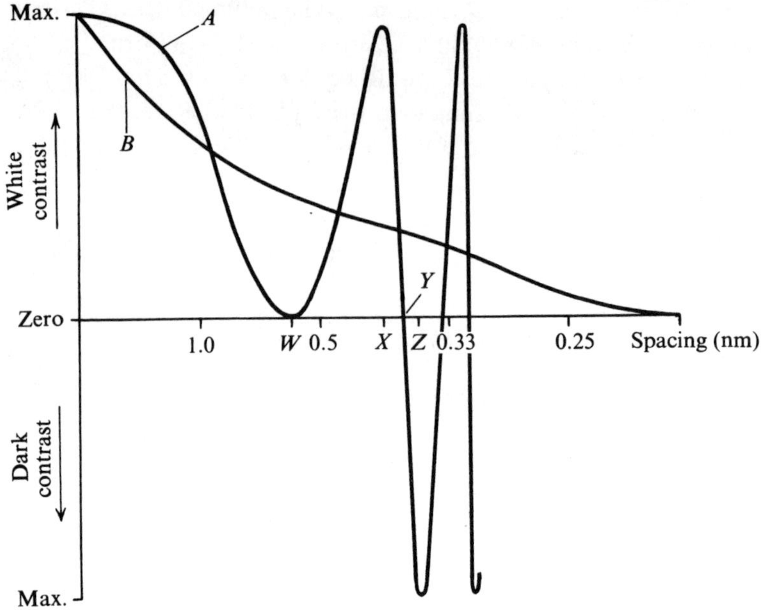

Fig. 5. The form of the amplitude contrast transfer function for two values of β_0/α_0. In curve A $\beta_0/\alpha_0 = 0$ and in curve B $\beta_0/\alpha_0 = 1$. For further explanation see text.

a high ratio of signal to noise. In TEM, however, the situation tends more towards $\beta_0/\alpha_0 = 0$ allowing the recording of phase contrast images with minimal beam damage. In considering in detail the importance of these parameters for high resolution and for a comparison of the relative performances of TEM and STEM instruments in bright-field and dark-field modes, as Burge (this volume) has done, resort must be made to the contrast transfer functions. These are derived from diffraction theory by Fourier transforms. To understand the relevance of these functions it is necessary to realise that any object can be split up into a series of sine and cosine components with wide-ranging periodicities. The contrast transfer functions tell us what fraction of each component present in the object is present in the final image and are usually plotted as a function of spatial frequency (reciprocal of the periodic spacings in the specimen). This principle is shown in Fig. 5 for amplitude contrast in which two extreme values of β_0/α_0 are shown, the case (A) for $\beta_0/\alpha_0 = 0$ is the situation to which TEM tends and (B) that for $\beta_0/\alpha_0 = 1$ which corresponds to a possible STEM condition. The precise values which these contrast envelopes assume are focus-dependent and in Fig. 5 are shown for zero defocus. Considering first condition A, a specimen periodicity of 1 nm would be imaged in medium white (positive) contrast.

A periodicity corresponding to *X* would have maximum contrast. Periodicities corresponding to *W* or *Y* would have zero contrast (i.e. information lost) and periodicities corresponding to *Z* would be imaged with maximum black contrast. For condition B, more similar to the STEM situation, the contrast remains positive and finite down to very small lattice spacings, making interpretation much simpler.

Heavy atom labelling of biological molecules

The effectively negligible chromatic defect of STEM and the ability to perform electronic signal processing (e.g. $N_e/N_i \propto Z$ mentioned above) allows detection of individual atoms of heavy elements, particularly favourably if they are contained in a low atomic number matrix such as biological tissue or other organic material.

The method shows great promise. Unstained DNA filaments and even non-negatively stained tRNA particles have been imaged (Wall & Langmore, 1970; Langmore & Wall, 1970) with the clear goal of eventually reading the genetic code by imaging DNA strands in which specific bases have been tagged by heavy metal atoms (Crewe, 1971). Progress has been made towards this exciting and important goal with the reported imaging of single mercury atoms in mercurated DNA strands both with STEM (Crewe, Langmore, Lamvik & Pullman, 1974) and TEM (Tanaka, Higashi-Fujime & Uyeda, 1975) instruments.

Burge shows that in STEM the relatively poor phase contrast contribution to bright-field imaging leads to a poor signal-to-noise ratio compared with TEM. Thus for a given detected signal the specimen sustains more radiation damage. The situation is reversed, however, for dark-field (DF) imaging, where STEM has very considerable DF advantages over TEM (e.g. DF images of unstained DNA filaments mentioned above). In TEM, methods of DF image formation, tilted beam or off-axis apertures or axial stop apertures all tend to introduce astigmatism, and further resolution and contrast loss is introduced by the imaging lenses; the STEM system avoids these drawbacks and the collection efficiency for elastically scattered electrons is very high. Further, without using an electron spectrometer, the ratio of elastically to inelastically scattered electrons in DF can be modified by careful selection of the detector aperture diameter giving effectively a DF image by elastic scattering only. A further great advantage is that dark-field and bright-field images can be recorded *simultaneously* in STEM by appropriate selection of detector channels to twin display screens.

Examination of radiation-sensitive specimens

The ability to record BF and DF images simultaneously, the great sensitivity of the electron detector system, and the ability to capture an image from a single scan and display it indefinitely on a high-resolution storage tube (several thousand line image) for examination and recording at leisure, mean that in practice STEM offers advantages in the examination of radiation-sensitive specimens.

Further, unlike the TEM, because the beam is focused to a small probe size there is no extraneous irradiation of areas other than that on which the spot impinges. Therefore, in radiation-sensitive specimens focusing can be accomplished on one area and the beam then moved to an adjacent area for image recording.

Conclusions

The high cost of these instruments alone means firstly that evaluation of the predictions and early results discussed above will be a slow process and secondly that STEM instruments will most probably never be anything like as common as TEM or SEM instruments. Nevertheless, it is expected that STEM instruments will make considerable contributions in several areas of microscopy in the future. Where the penetration of thick biological specimens (up to 0.5 μm or even 1 μm) is of prime importance the STEM (100 kV) may offer considerable advantage in resolution over the TEM at equivalent accelerating voltage. Although performance in these respects is unlikely to match that achievable with the 1 MeV high-voltage electron microscope, the lower cost, smaller size and greater ease of operation and maintenance may make it in practice the instrument of choice. In the field of high-resolution microscopy of ultrathin sections or individual macromolecules, while the ultimate resolution is no better than can be achieved with TEM, there are advantages in image brightness, freedom from rotation and ease of bright-field interpretation, and instrumental and specimen instabilities have less effect on the recorded image. The dark-field imaging mode is considerably more efficient in STEM than in TEM, leading to higher contrast in images of unstained biological molecules. The ease with which signal processing can be performed means that techniques such as Z discrimination can readily be applied, e.g. to visualisation of single heavy metal atoms in labelled biomolecules. The reduced contamination in the clean, ultra-high vacuum of the STEM and the ability to minimise radiation damage will probably be features of increasing importance in future high-resolution studies of biomolecules. Possibly the greatest advantage lies in the potential for microanalysis. Electron diffraction, X-ray energy dispersive microanalysis and electron energy loss spectroscopy can all be comfortably accommodated in the one instrument and in each mode the STEM offers advantages over the TEM. Therefore while the STEM will never rival the TEM for routine biological investigations we can confidently expect that in specialist centres it will make major contributions in the biological sciences.

Informative discussions with Drs J. Fryer and J. N. Chapman are gratefully acknowledged. VG Microscopes kindly provided evaluation time on an HB5 STEM instrument.

References

Crewe, A. V. (1971). A high-resolution scanning electron microscope. *Scientific American*, **224**, 26–35.

Crewe, A. V. (1974). Scanning transmission electron microscopy. *Journal of Microscopy*, **100**, 247–59.

Crewe, A. V. & Groves, T. (1974). Thick specimens in the CEM and STEM. I. Contrast. *Journal of Applied Physics*, **44**, 3662–72.

Crewe, A. V., Langmore, J. P., Lamvik, M. K. & Pullman, J. (1974). Scanning transmission electron microscopy of unstained and selectively stained biological specimens. *Electron Microscopy, Canberra*, **2**, 162–3.

Crewe, A. V. & Wall, J. (1970). A scanning microscope with 5 Å resolution. *Journal of Molecular Biology*, **48**, 375–93.

Crewe, A. V., Wall, J. & Langmore, J. (1970). Visibility of single atoms. *Science*, **168**, 1338–40.

Gabor, D. (1974). The history of the electron microscope, from ideas to achievements. *Electron Microscopy, Canberra*, **1**, 6–12.

Groves, T. (1975). Thick specimens in the CEM and TEM. Resolution and image formation. *Ultramicroscopy*, **1**, 15–31.

Langmore, J. & Wall, J. (1970). Conformational study of tRNA using the high resolution scanning electron microscope. *Procedeings of the 28th Annual Meeting of the Electron Microscopical Society of America, Houston, Texas*, ed. C. J. Arceneaux, pp. 254–55. Claitor's Publishing Division, Baton Rouge.

Tanaka, M., Higashi-Fujime, S. & Uyeda, R. (1975). Electron microscope images of mercury atoms bound to DNA filament. *Ultramicroscopy*, **1**, 7–14.

Wall, J. & Langmore, J. (1970). Studies of DNA using a 5 Å scanning microscope. *Proceedings of the 28th Annual Meeting of the Electron Microscopical Society of America, Houston, Texas*, ed. C. J. Arceneaux, pp. 252–3. Claitor's Publishing Division, Baton Rouge.

R. E. BURGE

Scanning transmission electron microscopy at high resolution

The scanning transmission electron microscope (STEM) introduced by Crewe (e.g. Crewe & Wall, 1970) has been extensively described from the instrumental point of view (see for references Crewe, 1974*a*, *b*). The requirement is familiar of the high brightness field emission gun and its associated vacuum system for high-resolution scanning transmission microscopy. The theory of high-resolution image formation in STEM has been examined and related to that in the conventional electron microscope (TEM) for various modes of bright- and dark-field operation (Zeitler & Thomson, 1970; Thomson, 1973; Langmore, Wall & Isaacson, 1973; Rose, 1974, 1975; Dekkers & de Lang, 1974; Engel, 1974; Engel, Wiggins & Woodruff, 1974; Howie, 1974). On the instrumental side, commercial development of high-resolution STEM has been stimulated both in the direction of purpose-built microscopes and towards field emission electron gun systems that may be attached, with contingent modifications, to a TEM (e.g. Someya *et al.*, 1974). Further, medium-resolution scanning transmission facilities (using a thermal emitter) may be made available by fitting a scanning attachment to a TEM (see e.g. Gentsch, Gilde & Reimer, 1974) or by modifying a TEM (e.g. Kuypers, Thompson & Andersen, 1973).

Commercial STEM instruments are being produced for electron energies in the 50–100 keV range. However, Crewe (see 1974*a*) is currently developing a 1 MeV scanning microscope directed towards a resolution of 0.1 nm (see also the high-voltage STEM of Cowley, 1970).

The applications of STEM to the imaging of biological specimens have so far been sparse and have principally concerned thin test specimens of use in evaluating the relative resolution on a point-to-point basis and electron beam damage between TEM and STEM. No results for quantitative image analysis in STEM have been published before the present article, which begins to explore the value of computer–STEM direct interaction.

The potential benefits of STEM in connection with thick specimens have been explored experimentally using polystyrene spheres by Gentsch *et al.* (1974), and theoretically by Crewe & Groves (1974), Reimer & Gentsch

(1975) and Groves (1975). Useful comments and references on the use of STEM for thick specimens in materials science are given by Fraser & Jones (1975) and Cowley (1975). For thick specimens STEM is in competition with TEMs having an electron beam energy in the 1 MeV range. Thick specimens will not be considered further here.

The use of STEM to produce energy-gated images by selecting only elastic electron scattering or a specific mode of inelastic scattering (energy gate width 1–2 eV) and as a high-resolution electron energy loss spectrometer (energy resolution ~ 0.3 eV) has been described by Crewe, Isaacson & Johnson (1971) (see also Egerton & Lyman, 1976; Browne, Lackovic & Burge, 1976). In connection with imaging by elastic scattering, STEM is in competition with the TEMs having an energy filter inserted within the column (see e.g. Henkelman & Ottensmeyer, 1974; Egerton, Philip & Whelan, 1976); however, the general problems of alignment and the use of the energy filter in TEM are considerably greater than in STEM. Henkelman & Ottensmeyer point out, in respect of bright-field TEM, the considerable gain in terms of increased image contrast and definition from separating the image formed by elastic scattering. In respect of the measurement of energy loss spectra (only) a simple electron spectrometer with an energy resolution of 1–2 eV can be attached for this purpose below the screen of the TEM (e.g. Cundy, Metherell & Whelan, 1966). For recent considerations of energy loss spectra and the possibilities of using STEM for microanalysis see Isaacson & Johnson (1975) and Ferrier (this volume).

In this article STEM, working in the 40–100 keV electron energy range, is assessed for examining biological specimens at high resolution (i.e. better than 1 nm) on the basis of about 18 months' experience with a Vacuum Generators HB5 STEM. The approach is comparative, assuming the availability of TEM. Both bright-field and dark-field imaging are considered. The present stage of development is shown in Fig. 1 of a system allowing the simultaneous digital recording of the bright-field image formed by elastically scattered electrons and the corresponding dark-field image. A measure of computer control of the sequence of operations, focus control, etc., is possible and being further developed.

The direct recording of images digitally does not by itself constitute an overwhelming advantage of STEM over TEM. Indeed, a similar financial investment to that required for the image-analysing facilities for STEM could be used to provide scanning and image analysis of photographic images taken on TEM.

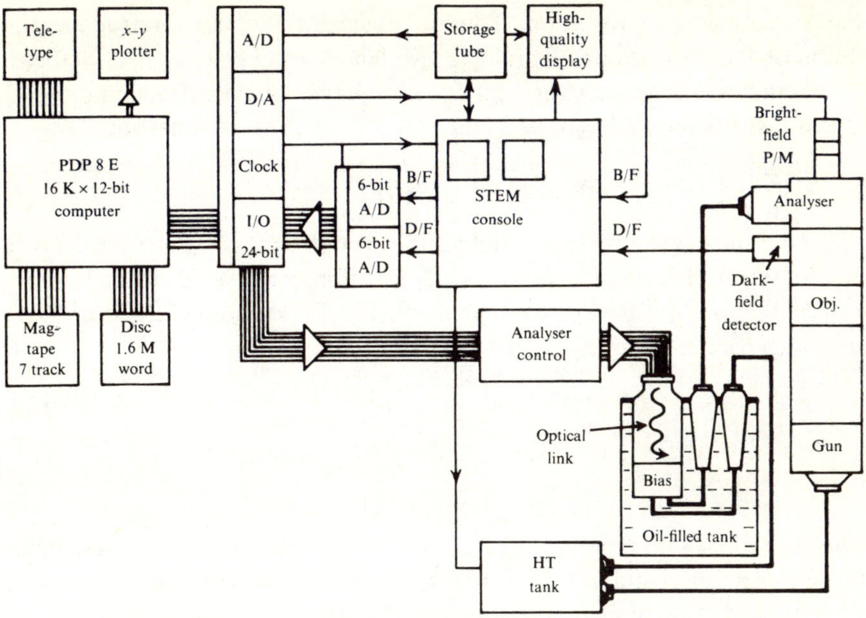

Fig. 1. Present state of development (1975) of data collection from and computer control of the Vacuum Generators HB5 STEM at Queen Elizabeth College.

High-resolution bright-field operation

TEM imaging

The strongest specimen–electron beam interaction consistent with resolution less than 1 nm is probably expressed by the weak phase/weak amplitude model (e.g. Erickson & Klug, 1970; Erickson, 1971). In this model, as referred to TEM, an incident parallel electron beam of unit amplitude (assumed to be totally spatially and temporally coherent), is transmitted by the specimen as the emergent wave function (object wave; $\Psi x,y$) corresponding to elastic scattering:

$$\Psi(x,y) = 1+iP(x,y)-A(x,y) \tag{1}$$

where $P(x,y)$ refers to the phase component of the object wave $\pi/2$ out of phase with the incident wave, and $A(x,y)$, the amplitude component, represents attenuation of the object wave from whatever cause.

The image intensity, neglecting noise, is given by

$$I(x,y) = 1-2P(x,y) * p(x,y)-2A(x,y) * a(x,y) \tag{2}$$

where $p(x,y)$ and $a(x,y)$ are the image point spread functions (i.e. correspond to images of object points) for phase and amplitude imaging respectively and * represents convolution. The point spread functions introduce the effects of the aberrations of the objective lens and its condition of defocus.

Taking the Fourier transform (or effectively, the optical diffraction pattern) of this distribution of intensity, gives in the absence of astigmatism:

$$\tilde{I}(\theta,\phi) = \delta(\theta,\phi)-2P(\theta,\phi)\sin\left[\frac{2\pi}{\lambda}\Delta\left(\frac{\theta}{\lambda}\right)\right]-2A(\theta,\phi)\cos\left[\frac{2\pi}{\lambda}\Delta\left(\frac{\theta}{\lambda}\right)\right], \quad (3)$$

where θ is the angle of scattering and ϕ the azimuthal angle. The tilde denotes the Fourier transform. The central delta function corresponds to the diffraction by the uniform background intensity. The Fourier transforms of $P(x,y)$ and $A(x,y)$ are respectively $P(\theta,\phi)$ and $A(\theta,\phi)$, while

$$\tilde{p}(\theta) = \sin\left[\frac{2\pi}{\lambda}\Delta\left(\frac{\theta}{\lambda}\right)\right], \quad (4a)$$

$$\tilde{a}(\theta) = \cos\left[\frac{2\pi}{\lambda}\Delta\left(\frac{\theta}{\lambda}\right)\right], \quad (4b)$$

where (4*a*) and (4*b*) denote the phase and amplitude contrast transfer functions respectively and Δ incorporates the effects of spherical aberration, defocus and the limited aperture, semi-angle α_0 of the objective lens $(0 \leqslant \theta \leqslant \alpha_0)$.

$$\Delta(\theta) = -\left(C_s\frac{\theta^4}{4}-\Delta f\frac{\theta^2}{2}\right), \quad (5)$$

(e.g. Erickson & Klug, 1970), where C_s is the spherical aberration coefficient of the objective lens and Δf the off-focus distance. Δf is taken to be positive for underfocus.

Taking the particular case of a weak phase object, we introduce, again for the TEM, the effects of partial spatial coherence (finite source size) and partial temporal coherence (finite thermal energy spread, 1–2 eV, of heated thermal filament).

Partial spatial coherence affects the diffraction pattern $\tilde{I}(\theta,\phi)$, under high-resolution conditions, as the envelope function $E(\theta)$ (Frank, 1973*a*) and partial temporal coherence (Hanszen & Trepte, 1971) affects it as the envelope $E'(\theta)$. Thus equation (3) becomes for the weak phase object:

$$\tilde{I}(\theta,\phi) = \delta(\theta,\phi)-2P(\theta,\phi)\,E(\theta)\,E^1(\theta)\sin\left[\frac{2\pi}{\lambda}\Delta\left(\frac{\theta}{\lambda}\right)\right]. \quad (6)$$

The effective phase contrast transfer functions for 80 keV electrons (C_s = 1.6 mm, corresponding to the Philips EM300 electron microscope) accounting for partial coherence are shown for various values of off-focus distance in Fig. 2. These curves correspond in TEM to practically realisable contrast transfer conditions for the weak phase object. The functions $E(\theta)$ and $E^1(\theta)$ also operate on the amplitude contrast transfer function, which,

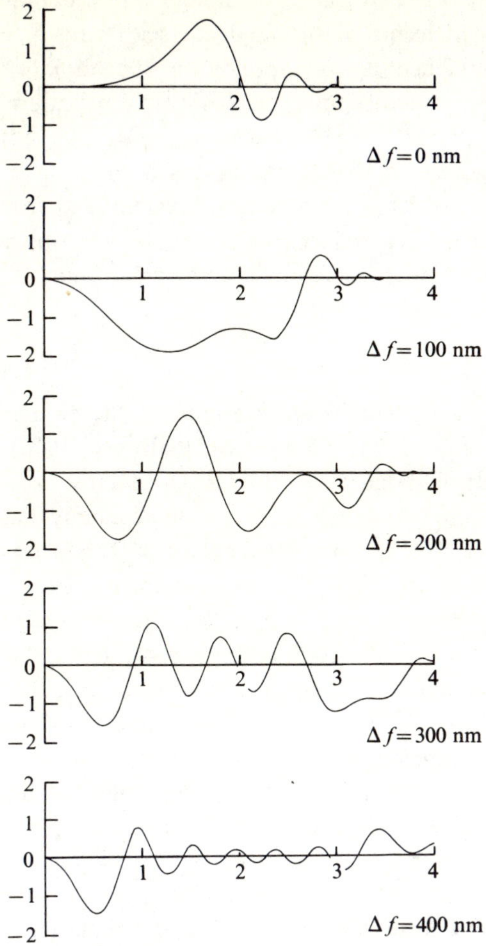

Fig. 2. Phase contrast transfer functions for high-resolution TEM. 80 keV electrons, $\lambda = 4.2 \times 10^{-3}$ nm, $C_s = 1.6$ mm. Partial spatial coherence characterised by $q_0 = 0.2\ \text{nm}^{-1}$ (Frank, 1973*a*) for Gaussian source distribution. Partial temporal coherence characterised by $\Delta E = 1$ eV (Hanszen & Trepte, 1971). Abscissa is $\theta/\lambda\ \text{nm}^{-1}$ (ordinate × 2).

of course, for a given value of Δf, has maxima at the angular positions where the phase contrast transfer function has zeros (see Fig. 4 for an example). The competing effects of spherical aberration and defocus on $E(\theta)$ (Frank, 1973*a*) produce a need, when considering detail at resolutions of 0.5 nm and better, to use defocus values (underfocus) greater than that corresponding to optimum defocus (i.e. where the phase contrast transfer function is flat over the widest range of spatial frequencies in the image) which occurs at $\Delta f = +82$ nm for the conditions of Fig. 2.

Fig. 2 prepresents the Fourier transform of an image which contains all spatial frequencies presented in terms of the angle of scattering θ, or the spatial frequency θ/λ nm^{-1}. Hence it corresponds to the radial profile measured from the central beam across an optical diffraction pattern. The sign changes of the phase contrast transfer function mean that adjacent frequency ranges in the image are rendered with reversed contrast and the frequencies corresponding to the zeros are irretrievably removed from the image. These zeros correspond to the well known minima in the optical diffraction pattern from a high-resolution image of a thin carbon film.

STEM imaging

In respect of STEM, and elastic scattering, we invoke the principle of reciprocity (Zeitler & Thomson, 1970; Howie, 1974; Engel, 1974). Ray diagrams comparing the bright-field operation of STEM and TEM are shown in Fig. 3. The equivalence of these two modes on inversion of the ray diagrams was first pointed out by Zeitler & Thomson; the detector in STEM takes the place of the illuminating system in TEM. The equivalence is exact when $\beta_0/\alpha_0 \ll 1$, with β_0, the illumination or detection angle, corresponding to the usual conditions in TEM ($\beta_0/\alpha_0 \sim 10^{-4}/10^{-2}$ at high resolution); but this implies in STEM a very small angle of collection compared with the beam angle and hence a highly inefficient electron collecting system in bright-field.

For $\beta_0/\alpha_0 \ll 1$, equation (6) corresponds also to the Fourier transform of a STEM bright-field micrograph of a phase object. We consider first the envelope functions $E(\theta)$ and $E^1(\theta)$. It is convenient to use the experimental results of Someya *et al.* (1974) in respect of partial spatial coherence. From their results the coherence length for $\alpha_0 = 10^{-2}$ rad lies between values of 20 nm and 35 nm (the published results have $\alpha_0 > 3\times10^{-3}$ rad). Thus from these results $E(\theta)$ may be taken as effectively unity for high-resolution studies in STEM. In contrast, for the HB5, under normal conditions of condenser operation, there is evidence that the limit to high-resolution performance is set by the limited partial spatial coherence of the probe. In relation to partial temporal coherence, the energy width ΔE is less than 0.3 eV at half height of the energy distribution for the field emitter of the HB5 under high-resolution conditions (total beam current $\sim$ 10 μA); hence $E^1(\theta)$ may be taken to be unity for all potential purposes with STEM.

Thus, there should be a higher degree of freedom in STEM for choosing the transfer function, and hence the defocus, and a reasonable choice to examine is the so-called balanced condition, which corresponds to the condition for optimum defocus in TEM, where the illuminating angle α_0 (see equation (5)) is given by $C\,\alpha_0^4/4 = \lambda = \Delta f\alpha_0^2/2$.

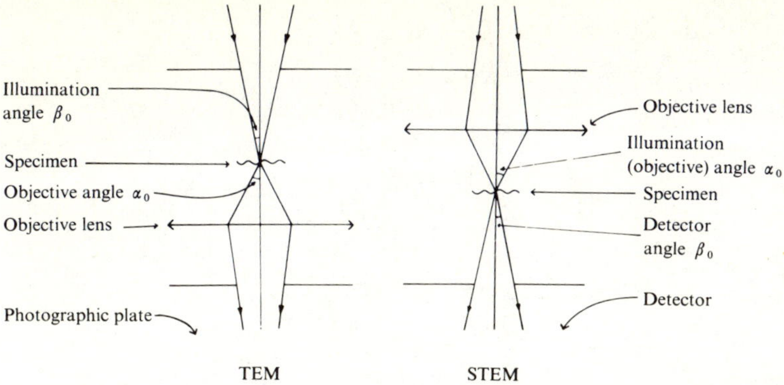

Fig. 3. Comparison of ray diagrams for bright-field TEM and bright-field STEM. The objective aperture semi-angle (TEM) and the probe convergence angle (STEM), α_0, and the illumination angle (TEM) and detector angle (STEM), β_0, correspond.

i.e. the spherical aberration is balanced by the defocus at an angle $\theta = \alpha_0$ which defines the scanning probe. The corresponding values for 80 keV electrons and $C_s = 1.6$ mm (assuming the same objective lens in STEM as TEM) are $\alpha_0 \sim 10^{-2}$ rad, $\Delta f = +82$ nm (underfocus).

It is unrealistic from the point of view of beam damage to operate STEM with $\beta_0/\alpha_0 \ll 1$. Further, as $\beta_0 \to \alpha_0$ (with a single detector, see below) the phase contrast progressively decreases as a greater fraction of both the incident beam and the elastically scattered electrons is collected. The changes in both the phase and amplitude contrast transfer functions are shown in Fig. 4 for $0 \leqslant \beta_0/\alpha_0 \leqslant 1$ (modified from Thomson, 1973), $\alpha_0 \simeq 10^{-2}$ rad, $\Delta f = +82$ nm.

For $\beta_0/\alpha_0 \ll 1$, the maximum angle through which collected electrons are scattered is α. As the collection angle increases the maximum scattering angle for collection also increases up to $2\alpha_0$ when $\beta_0/\alpha_0 = 1$. Obviously the phase contrast reduces dramatically as the electron collecting power of the detector increases, $\beta_0/\alpha_0 \sim 0.2$ giving reasonable contrast for high-resolution studies. The effect of change of β_0/α_0 on the amplitude contrast transfer function is not so dramatic. The maximum spatial frequency in a phase image with good contrast is $\alpha_0/\lambda = 0.42$ nm^{-1} at the balanced condition at 80 keV (this resolution will improve for a smaller C_s, as may be produced for a specially designed objective lens rather than a commercially available lens). Thus the probe convergence angle represents a more fundamental parameter than the objective aperture in the TEM which may be varied or removed altogether for high-resolution images by phase contrast. Put another way, to produce a higher resolution with high phase contrast in the given example,

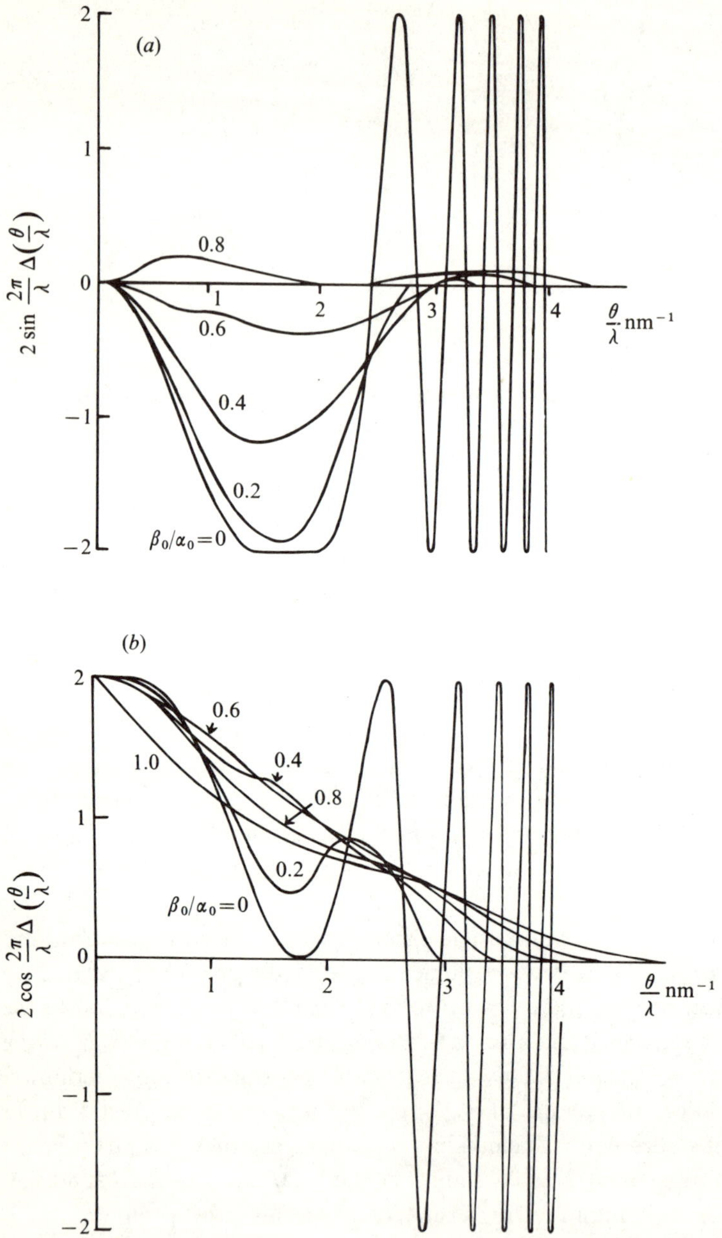

Fig. 4. Comparison of bright-field phase contrast transfer functions for optimum defocus conditions ($\Delta f = +82$ nm, 80 keV electrons, $C_s = 1.6$ mm) and various ratios β_0/α_0. The transfer function for bright-field TEM, with normal parallel illumination, corresponds to $\beta_0/\alpha_0 = 0$ and (as shown) extends beyond the angle α_0 ($\equiv 2.4$ nm^{-1})

will require a larger beam aperture than is consistent with the balanced condition.

Thomson (1973) has shown that the maximum signal-to-noise ratio for a given signal in STEM occurs for $\beta_0/\alpha_0 = 0.45$ where phase contrast is very significantly lower than for small β_0/α_0 (Fig. 4). For $\beta_0/\alpha_0 \sim 0.2$ in STEM, for high-contrast imaging, the relative signal-to-noise ratios for TEM (axial illumination) compared with STEM lie between 5:1 and 6:1 for the imaging of single atoms, when making use of an axially centred single detector aperture. The validity of working with $\beta_0/\alpha_0 \sim 0.5$ is dependent upon the discrimination between contrast levels attainable in the recording system; measurements with such large detector apertures have not yet been reported.

Rose (1974) has shown how the electron collection efficiency in bright-field STEM may be improved by the device of a central circular detector aperture and two annular detectors such that the total angle of collection β_0 is equal to α_0. The signal from the inner annulus is subtracted from the sum of the signals from the central detector and the outer annulus.

The principle of the Rose method is made clear from Fig. 5 which shows the phase contrast transfer function for 80 keV electrons ($C_s = 1.6$ mm) derived from the imaging conditions which were established in general terms by Rose. A particular setting of off-focus distance Δf^1 is required, where Δf^1 is 1.44 times the value for the balanced condition. The transfer function is the nearest approach to a step function, shown as a dashed line in Fig. 5, with the three areas of beam collection separated by angles β_1, β_2 as marked. Rose shows that the highest signal-to-noise ratio for a given input signal occurs when $\beta_1/\alpha_0 = 0.51$, $\beta_2/\alpha_0 = 0.87$, the central collector angle being similar to that defined by Thomson (1973) for a single collector. The optimum beam angle α_0 is also marked, corresponding to a resolution of 0.3 nm. The subtraction of signals accounts for the sign changes between adjacent loops of the transfer function.

It might be anticipated from Fig. 5 that, since all the electrons within the incident cone α_0 at the spectrum were collected, the image-forming conditions would be identical to those in bright-field TEM. However, Fig. 5 corresponds to the distribution of electrons at the detector in bright-field phase contrast imaging arising from a *pencil* of electrons incident perpendicular to the specimen. An integration must be performed over the cone α_0 of illumination

---

which is the cut-off point for the scanning probe for $\beta_0/\alpha_0 \ll 1$. For ratios $\beta_0/\alpha_0 \neq 0$, the curves for $\theta/\lambda > \alpha_0$ correspond to the STEM case only. Envelope functions for partial spatial and temporal coherence have not been used. (*b*) As (*a*) in respect of both TEM and STEM details but showing the amplitude contrast transfer function. STEM beam half-angle $\alpha_0 \equiv 2.4\ \text{nm}^{-1}$.

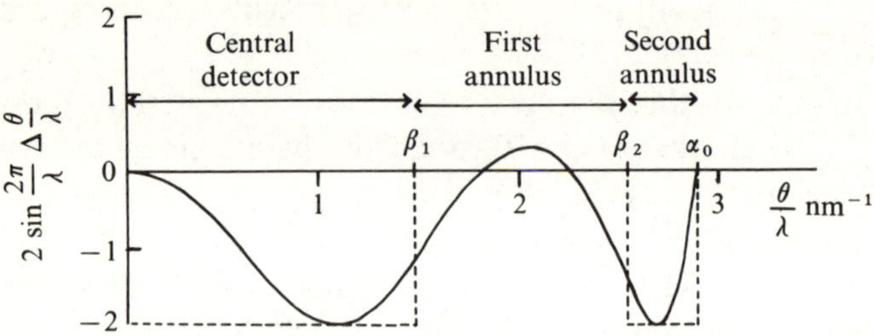

Fig. 5. STEM phase contrast transfer function with Rose (1974) annular detection. The curve as drawn refers to $\beta_0/\alpha_0 = 0$ where the transfer function approximates a step function. For the real case, $\beta_0/\alpha_0 = 1$, integration is necessary over the cone of the incident illumination. Optimised defocus value $\Delta f = \Delta f^1 = +118$ nm. STEM beam half-angle $\alpha_0 \equiv 2.9 \text{ nm}^{-1}$.

which involves a degree of cancellation of the positive and negative loops of the transfer function.

With an annular detection system, optimised as described above, Rose shows that for a theoretical point object, the ratio of the number of incident electrons in STEM to the number incident for TEM with parallel illumination for the same signal-to-noise ratio is 3:1 to 4:1. It is unlikely that much further improvement can be expected in this ratio by alterations of the detector geometry, although the possibility of a zone plate in the STEM illuminating system might be explored (Rose, 1974).

High-resolution dark-field operation

A comprehensive study of single atom contrast in TEM in dark-field has appeared recently (Chiu & Glaeser, 1975) which is complementary to the comparative discussion of the dark-field performance of STEM and TEM of Langmore *et al.* (1973).

In TEM, dark-field contrast arises by suppressing the unscattered beam and allowing only the scattered electrons to pass through a finite objective aperture. In STEM, the electrons scattered outside the incident beam defined by α_0 may be collected on a large aperture annular detector. The electron detecting efficiency of STEM in this mode is thus greater by a factor of 2 to 4 than that of TEM under optimum conditions for all elements relevant to unstained or stained biological material; the increased efficiency of STEM for the detection of heavy atoms is particularly striking (see figs. 3 and 4 of Langmore *et al.*, 1973). Similarly, according to the calculations of Thomson (1975), the relative signal-to-noise ratio for STEM:TEM dark-field, for a given number of electrons incident on the specimen, is about 2:1 both for single carbon and single gold atoms; thus for the same signal-to-noise ratio and these two elements, four times the number of electrons incident on the specimen are needed for TEM as against STEM. Thomson (1975) further shows that for a fixed number of electrons incident on the specimen, the maximum signal-to-noise ratio achievable in dark-field STEM is approximately the same as the best signal-to-noise ratio for bright-field TEM for the same number of incident electrons. The same spatial resolution may be expected for the two instruments for the same α_0.

Considering the interpretation of dark-field images in terms of the weak phase/weak amplitude object model, we find for TEM that the intensity distribution and its Fourier transform are complicated. In contrast, on the theoretical basis, validated by the calculations of Engel *et al.* (1974), that the hole in the centre of the annular detector for dark-field STEM is small (Misell, Stroke & Halioua, 1974) we may write for the distribution of current at the image point, $r = (x,y)$,

$$I(r) = \int [P^2(r^1)+A^2(r^1)]|\zeta(r-r^1)|^2 dr^1 \qquad (7)$$

where $|\zeta(r)|^2$ represents the squared modulus of the probe transmission function. This gives a method for determining $[P^2(r)+A^2(r)]$ or, of course, for a weak phase object, a method for $P(r)$.

A comparative view of STEM and TEM for image processing

Bright-field and dark-field image pairs for comparable specimens mounted on thin carbon films in TEM (Philips EM300) and STEM (Vacuum Generators HB5) are compared in Fig. 6. (The specimen consists of a negatively stained preparation of spine-like appendages from a Gram-negative bacterium of the family Pseudomonadaceae (McGregor-Shaw, Easterbrook & McBride, 1973); the striations are 11.5 nm apart as determined by X-ray diffraction.) It is not intended to imply that a negatively stained specimen is a suitable vehicle for high-resolution image processing. The bacterial spines provide a high-contrast image useful for the present comparative purpose; the effective high-resolution specimen is the background film.

The STEM photographs were produced by converting the electron current as the electron probe scans the specimen linearly into digital signals recorded via the PDP8 computer on magnetic tape. The bright-field signals were recorded using a single axial detector with $\beta_0/\alpha_0 \sim 0.2$, while the dark-field detector covered a large angular range from $\alpha_0 \sim 10^{-2}$ rad. The independent bright-field and dark-field signals are recorded *simultaneously* (not

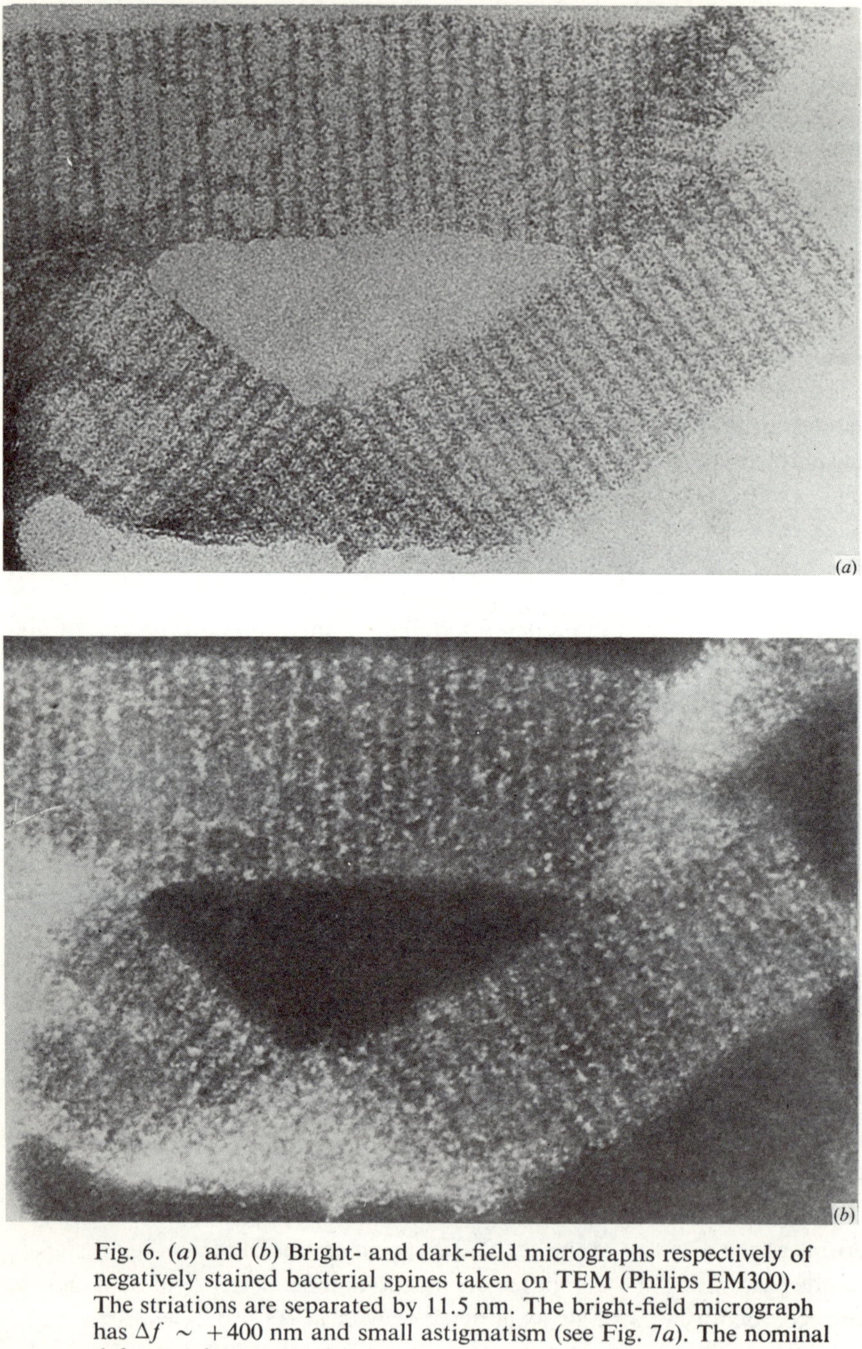

Fig. 6. (*a*) and (*b*) Bright- and dark-field micrographs respectively of negatively stained bacterial spines taken on TEM (Philips EM300). The striations are separated by 11.5 nm. The bright-field micrograph has $\Delta f \sim +400$ nm and small astigmatism (see Fig. 7*a*). The nominal defocus value was unchanged for the dark-field photograph.

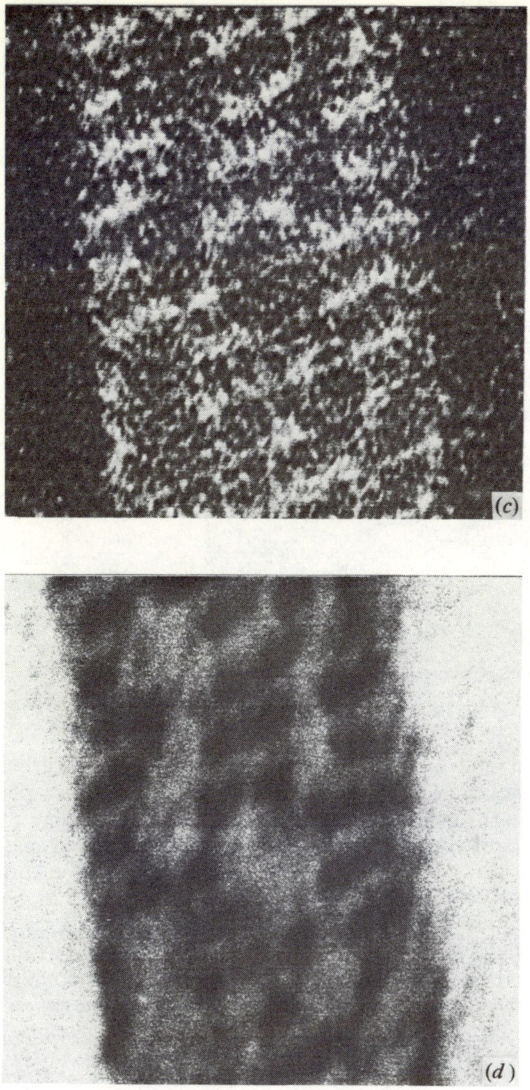

(*c*) and (*d*) A corresponding pair of bright-field ($\beta_0/\alpha_0 \sim 0.2$) and dark-field images taken *simultaneously* using the STEM HB5 and recorded digitally on magnetic tape. There are 512×512 picture points for the area shown. The magnetic tape was replayed through a computer microfilm plotter device and photographed with calibration grey levels. ((*c*) and (*d*) are reversed in contrast compared with (*a*) and (*b*).)

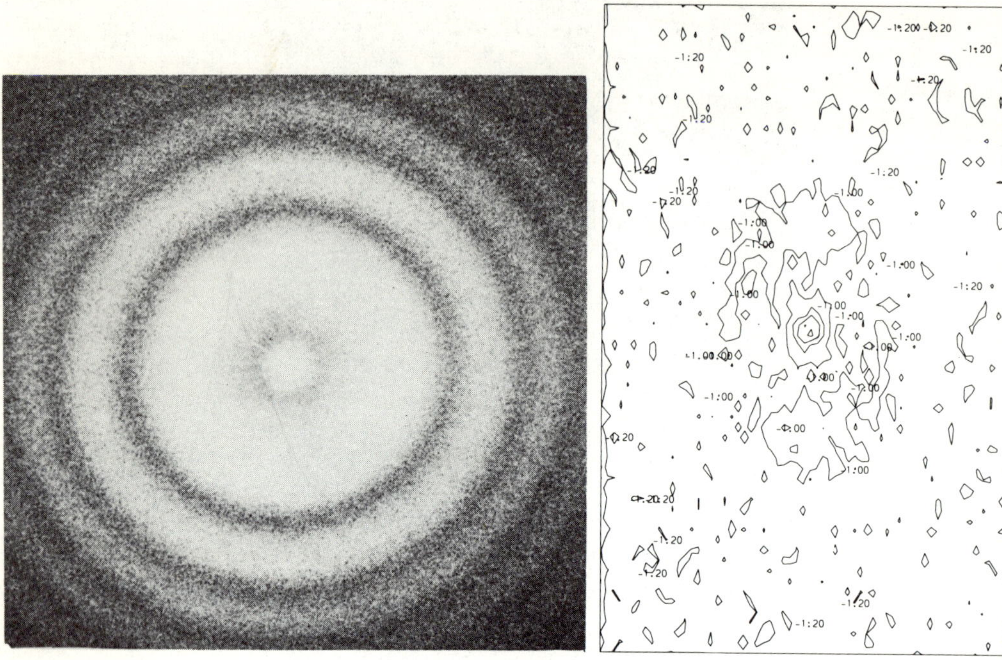

(*a*) (*b*)

Fig. 7. (*a*) Optical diffraction pattern of background carbon film corresponding to Fig. 6(*a*). Note the many zeros of the phase contrast transfer function for the large defocus value (+400 nm underfocus). The micrograph exhibits negligible astigmatism (when small astigmatism is present the transfer function becomes elliptical rather than circular). (*b*) Contour plot of calculated Fourier transform from a STEM HB5 bright-field migraph of carbon film ($\beta_0/\alpha_0 \sim 0.2$), near focus and showing clearly only a single broad ring at about 1 nm^{-1} of the phase contrast transfer function arising from an astigmatic probe (visual correction) with limited spatial coherence observed near focus ($\Delta f \sim 0$; see corresponding TEM case in Fig. 2). The maximum spatial frequency represented is 2 nm^{-1}.

sequentially and thus subject to both alteration of the imaging system and additional beam damage as in TEM). The dark-field TEM image was produced by the tilted beam method.

The phase contrast transfer function for the bright-field TEM picture obtained by optical diffraction from the background image of the carbon film is shown in Fig. 7(*a*). The original film shows seven zeros of the transfer function (see Fig. 2). Results for the phase contrast transfer function for a thin carbon film in STEM are shown in Fig. 7(*b*). Optical diffraction of the carbon film image in dark-field TEM shows only a monotonic fall off in intensity of superimposed elastically and inelastically scattered electrons from the centre (corresponding to the frequency of occurrence for large spacings) to the minimum recorded spacings. Finally, the calculated diffraction from the dark-field STEM images of thin carbon films shows essentially the autocorrelation function of the scanning probe of a size determined by diffraction at the limiting probe aperture. Further evaluation of these Fourier transforms is required in connection with the possible effects of partial spatial coherence with a highly convergent probe and of the quantitative validity of the assumption of incoherent dark-field imaging (equation (7)). Such detailed studies of STEM imaging are essential to image processing, as will appear below in relation to STEM applications.

The theory of bright-field imaging of high-resolution objects in the TEM is fully developed and has given rise to methods of determining the phase and amplitude distributions of the image (e.g. for the phase/amplitude object and use of two micrographs of different defocus in order to determine $P(x,y)$ and $A(x,y)$ (Schiske, 1968; Frank 1972, 1973*b*; Burge & Scott 1975*a*) or, for the weak phase object, the determination of $P(x,y)$ by optical methods (Stroke & Halioua, 1973; Stroke, Halioua, Thon & Willasch, 1974; Burge & Scott, 1975*b*, 1976)). It has become clear that because of the limited spatial coherence (perhaps a spatial coherence 'patch' 3–6 nm in diameter at the specimen for the normal high-resolution conditions in TEM) and temporal coherence of the electron emission from a thermal filament, defocus distances of +300 nm to 400 nm (underfocus) are required (see Fig. 2) in order to optimise the high-frequency image components. This degree of defocus is rather close to that for the optimum resolution of the image due to inelastically scattered electrons superimposed on the elastic image, and separation of the elastic image by energy analysis is essential to take full advantage of the benefits provided by the high spatial coherence. For weak phase/weak amplitude objects, high defocus implies many zeros of the transfer function with corresponding loss of information at particular spatial frequencies.

Even with an optimised detection system (Rose, 1973), with the implied limitation of image contrast and limitation of the off-focus distance to a particular value, the STEM cannot compete with the TEM because of beam damage for bright-field phase contrast imaging alone. However, the STEM has operating advantages of which the most important, assuming the recording by energy analysis of the bright-field elastic image, is the simultaneous recording of bright- and dark-field elastic images. There are no effects due to chromatic aberration in STEM even without energy analysis, as the electrons are focused by the objective lens onto the specimen. However, it may be that the image-blurring effect of delocalisation (of position in the specimen (e.g. Isaacson, Langmore & Rose, 1974) where the inelastic event may be regarded as occurring) will act in STEM in a similar fashion to the chromatic effect. When energy analysis is available, the inelastic STEM signal, which is confined to small angles of scattering, may also be recorded simultaneously with the elastic images, the level of specimen irradiation being governed by that required for the bright-field elastic image. This gives various possibilities for signal processing, e.g. the ratio of the elastic to inelastic images and their differences, as discussed by Crewe, Langmore & Isaacson (1975). The inelastic image by itself is not useful for high-resolution work.

The STEM has as clear an advantage for dark-field imaging as TEM has for bright-field. Just as optical and computer correction for the transfer functions are readily applied to bright-field TEM, so assuming incoherent imaging, correction of the STEM dark-field images for the probe size and current disbution is readily accomplished optically (Stroke & Halioua, 1972) or by computation. Again, the phase and amplitude components of the image occur for dark-field STEM in a simple linear form, as they do in bright-field TEM, giving the possibility of their separate determination from an appropriate pair of images.

STEM applications at high resolution

The future of high-resolution electron microscopy is bound up with methods of minimising beam damage and of image interpretation. It is a happy accident that the weak phase/weak amplitude object in both bright-field TEM and dark-field STEM is examined under minimal beam damage conditions for the mode, and, also, in principle at least, in a simple situation for interpretation. Of course, signal-to-noise optimisation must also be considered.

Three particular features of the STEM in bright-field should be pointed out: (i) the dominance of the amplitude contrast transfer function (Fig. 4) over the phase contrast transfer function as β_0/α_0 increases towards unity; (ii) the freedom of the amplitude contrast transfer function from gaps in the presentation of image spatial frequencies; (iii) the possibility of making use of the focus-dependence of the phase contrast image to define structural features in the z dimension perpendicular to the x, y specimen plane.

As $\beta_0/\alpha_0 \rightarrow 1$, the amplitude transfer function is a monotonically decreasing function of spatial frequency. The importance of the amplitude image for the detection of heavy atoms in the presence of light ones has been pointed out by Frank (1972, 1973*b*) in respect of TEM imaging. The point at issue is the ratio of the real to the imaginary part of the atomic electron scattering factors for $0 \leqslant \theta \leqslant \alpha_0$ for heavy and light atoms. The real part f_0 contributes to the phase image and the imaginary part f^1 to the amplitude image. The tabulations show that for 80 keV energy electrons e.g.

$$f_0(\mathrm{Hg})/f_0(\mathrm{C}) \sim 4, f^1(\mathrm{Hg})/f^1(\mathrm{C}) \sim 28,$$

assuming a mercury-labelled carbon matrix. Thus STEM bright-field imaging is highly favourable for the determination of the sites of specific heavy atom labels to biological molecules projected onto the x,y plane.

The z dimension of a heavy atom site is difficult to determine in TEM, although tilting the object plane may assist. In STEM the variation of atomic phase contrast with defocus distance has been considered by Rose (1974) for the location of single atoms in specimens of thickness t, where $t < \lambda_E$, the mean free path for elastic electron scattering ($\lambda_E \sim 110$ nm for carbon and 80 keV energy electrons). Using an optimised bright-field annular detector, the intensity of a phase contrast atomic image decreases rapidly with change of focus from the optimum defocus value. Rose shows the change is proportional to $(\Delta f)^2$ as against proportional to Δf for TEM. These changes are shown in Fig. 8 for point objects (in the first Born approximation, Wentzel potential). In practice the complex nature of atomic scattering should be accounted for which produces an amplitude as well as a phase component of increasing importance for increasing atomic number. In principle, the z coordinate of a heavy atom could be established in a moderately thick specimen with errors related, according to the accuracy of measurement, to the 50 nm half-width of the central peak in Fig. 8.

Because of the favourable image conditions in dark-field STEM, we consider the function $[P^2(r)+A^2(r)]$, equation (7), for a weak phase/weak amplitude object. More generally, the same possibility arises of determining $|\Psi(x,y)|^2$ for the general object wave $\Psi(x,y)$ irrespective of the nature of the specimen–electron beam interation. Such results are affected by the usual problems of dark-field imaging, e.g. a weak phase object with a periodic sinusoidal distribution of phase of spatial frequency f, will be imaged as a sinusoid of frequency $2f$. However, image interpretation will be assisted by the *simultaneous* recording of the bright- and dark-field images. Indeed, for the weak phase/weak amplitude object in bright-field for $\beta_0/\alpha_0 \rightarrow 1$, the amplitude image $A(r)$ is recoverable while $[P^2(r)+A^2(r)]$ may be determined from dark-field measurements. These measurements provide a simple method for the complete solution of the object wave, in this case with the beam damage of a single exposure. Thus is provided the maximum possible information about a high-resolution object as affected by noise for assessment at the limit of resolution (< 0.3 nm at 80 keV with the optimised probe and detection system).

The possibility of imaging unstained biological specimens in dark-field is of importance. Put simply, if there is point in such measurements in dark-

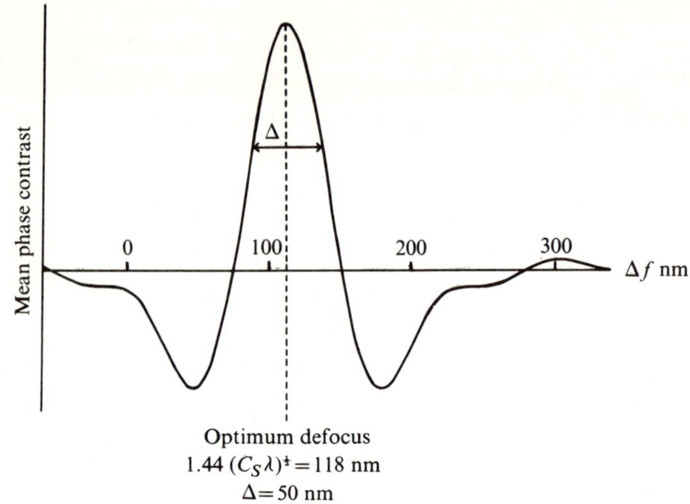

Fig. 8. Mean phase contrast corresponding to the radial and azimuthal integration of the transfer function for point object as a function of underfocus. The optimised underfocus value is $\Delta f^1 = 1.44\,(C_s\lambda)^{\frac{1}{2}} = +118$ nm. (Adapted from Rose (1974) for $C_s = 1.6$ mm, 80 keV electrons.)

field TEM there will be gains of two kinds in STEM. The first is reduced beam damage (i.e. higher detected signal) for a given number of incident electrons. The second is the high efficiency of detection of elastically scattered electrons irrespective of atomic number. The scattering increases with atomic number, making dark-field STEM especially favourable for the detection of heavy atoms.

In respect of suitable specimens for STEM, account must be taken of the possibility of specimen-carried contamination (despite the specimen stage pressure of $\sim 10^{-9}$ torr for the HB5) which, because of the high probe current, may reach as high a contamination rate as 10^5 nm s^{-1} (Dubochet, 1975). This may be alleviated by baking the specimen for an hour or so, within the microscope, at a temperature of 50–100 °C.

Other STEM applications

This article has concentrated on high-resolution imaging. The potential of STEM also embraces: (i) thick specimens and competition with high-voltage electron microscopes; (ii) electron loss and X-ray microanalysis within the context of an efficient system of data retrieval; and (iii) the use of the on-line system of data analysis made possible by the STEM scanning probe with pattern recognition techniques leading to computer control of some projects in image analysis. Also, the computer control of focus and of recording a given sequence of bright-field, dark-field and energy loss images followed by the taking of the energy loss spectra for identified specimen regions, each as small as about 1–2 nm in diameter, will soon be a routine procedure.

It is a pleasure to acknowledge the assistance of the optical and electron optical groups within the laboratory. The work is supported by the Science Research Council.

References

Browne, M. T., Lackovic, S. & Burge, R. E. (1976). Instrumentation and recording for the Vacuum Generators HB5 STEM instrument. In *Proceedings of the Institute of Physics, EMAG Meeting*, ed. J. A. Venables, pp. 27–30. Academic Press, New York & London.

Burge, R. E. & Scott, R. F. (1975*a*). Optical filtering for phase and amplitude determination in full bright field electron microscopy. *Optik*, **43**, 301–5.

Burge, R. E. & Scott, R. F. (1975*b*). Binary filters for high-resolution electron microscopy. I. *Optik*, **43**, 53–64.

Burge, R. E. & Scott, R. F. (1976). Binary filters for high-resolution electron microscopy. II. *Optik*, **44**, 159–72.

Chiu, W. & Glaeser, R. M. (1975). Single atom image contrast: conventional dark-field and bright-field electron microscopy. *Journal of Microscopy*, **103**, 33–61.

Cowley, J. M. (1970). High-voltage transmission scanning electron microscope. *Journal of Applied Crystallography*, **3**, 49–58.

Cowley, J. M. (1975). A comparison of scanning and fixed beam high-voltage electron microscopy. In *Physical Aspects of Electron Microscopy and Microprobe analysis*, ed. B. M. Siegel & D. R. Beaman, pp. 17–28. Wiley, New York.

Crewe, A. V. (1974*a*). Scanning transmission electron microscopy. *Journal of Microscopy*, **100**, 247–59.

Crewe, A. V. (1974*b*). Scanning microscopes: the approach to 1 Å. In *Quantitative Scanning Electron Microscopy*, ed. D. B. Holt, M. D. Muir, P. R. Grant & I. M. Boswarva, pp. 65–72. Academic Press, New York & London.

Crewe, A. V. & Groves, T. (1974). Thick specimens in the CEM & STEM. I. Contrast. *Journal of Applied Physics*, **45**, 3662–72.

Crewe, A. V. Isaacson, M. & Johnson, D. (1971). A high-resolution electron spectrometer for use in transmission scanning electron microscopy. *Review of Scientific Instruments*, **42**, 411–20.

Crewe, A. V., Langmore, J. P. & Isaacson, M. S. (1975). Resolution and contrast in the scanning transmission electron microscope. In *Physical Aspects of Electron Microscopy and Microbeam Analysis*, ed. B. M. Siegel & D. R. Beaman, pp. 47–62. Wiley, New York.

Crewe, A. V. & Wall, J. (1970). A scanning microscope with 5 Å resolution. *Journal of Molecular Biology*, **48**, 375–93.

Cundy, S. L., Metherell, A. J. F. & Whelan, M. J. (1966). An energy analysing electron microscope. *Journal of Scientific Instruments*, **43**, 712–15.

Dekkers, N. H. & de Lang, H. (1974). Differential phase contrast in a STEM. *Optik*, **41**, 452–6.

Dubochet, J. (1975). Carbon loss during irradiation of T4 bacteriophages and *E. coli* bacteria in electron microscopes. *Journal of Ultrastructure Research*, **52**, 276–88.

Egerton, R. F. & Lyman, C. E. (1976). An electron spectrometer system for use with a scanning transmission electron microscope. In *Proceedings of the Institute of Physics, EMAG meeting*, ed. J. A. Venables, pp. 35–8. Academic Press, New York & London.

Egerton, R. F., Philip, J. G. & Whelan, M. J. (1976). Applications of energy analysis in a transmission electron microscope. In *Proceedings of the Institute of Physics, EMAG meeting*, ed. J. A. Venables, pp. 137–40. Academic Press, New York & London.

Engel, A. (1974). The principle of reciprocity and its application to conventional and scanning dark field electron microscopy. *Optik*, **41**, 117–26.

Engel, E., Wiggins, J. W. & Woodruff, D. C. (1974). A comparison of calculated images generated by six modes of transmission electron microscopy. *Journal of Applied Physics*, **45**, 2739–47.

Erickson, H. P. (1971). The Fourier transform of an electron micrograph – first order and second order theory of image formation. In *Advances in Optical and Electron Microscopy*, vol. 4, ed. R. Barer & V. E. Cosslett, pp. 163–99. Academic Press, New York & London.

Erickson, H. P. & Klug, A. (1970). The Fourier transform of an electron micrograph: effects of defocusing and aberrations, and the implications for the use of underfocus contrast enhancement. *Deutschen Bunsen–Gesellschaft für Physikalische Chemie*, **74**, 1129–37.

Frank, J. (1972). A study of heavy/light atom discrimination in bright-field electron microscopy using the computer. *Biophysical Journal*, **12**, 484–511.

Frank, J. (1973*a*). The envelope of electron microscope transfer functions for partially coherent illumination. *Optik*, **38**, 519–36.

Frank, J. (1973*b*). Use of anomalous scattering for element differentiation. In *Image Processing and Computer-aided Design in Electron Optics*, ed. P. W. Hawkes, pp. 196–211. Academic Press, New York & London.

Fraser, H. L. & Jones, I. P. (1975). A note on the increase in usable foil thickness in scanning transmission electron microscopy. *Philosophical Magazine*, **31**, 225–8.

Gentsch, P., Gilde, H. & Reimer, L. (1974). Measurement of the top bottom effect in scanning transmission electron microscopy of thick amorphous specimens. *Journal of Microscopy*, **100**, 81–92.

Groves, T. (1975). Thick specimens in the CEM and STEM. Resolution and image formation. *Ultramicroscopy*, **1**, 15–31.

Hanszen, K. J. & Trepte, L. (1971). Der Einfluss von Strom- und Spannungsschwankungen sowie der Energiebriete der Strahlelektronen auf Kontrastubertragung und Auflösung des Elektronenmikroskops. *Optik*, **32**, 519–38.

Henkelman, R. M. & Ottensmeyer, F. P. (1974). An energy filter for biological electron microscopy. *Journal of Microscopy*, **102**, 79–94.

Howie, A. (1974). Theory of diffraction contrast effects in the scanning electron microscope. In *Quantitative Scanning Electron Microscopy*, ed. D. B.

Holt, M. D. Muir, P. R. Grant & I. M. Boswarva, pp. 183–209. Academic Press, New York & London.

Isaacson, M. S. & Johnson, D. (1975). The microanalysis of light elements using transmitted electrons. *Ultramicroscopy*, **1**, 33-52.

Isaacson, M. S., Langmore, J. P. & Rose, H. (1974). Determination of the non-localisation of the inelastic scattering of electrons by electron microscopy. *Optik*, **41**, 92–6.

Kuypers, W., Thompson, M. N. & Andersen, W. H. J. (1973). Scanning electron microscopy (1973 (Part 1)). In *Proceedings of* 6*th Annual Scanning Electron Microscope Symposium*, Chicago, pp. 9–16. IIT Institute, Chicago.

Langmore, J. P., Wall, J. & Isaacson, M. S. (1973). The collection of scattered electrons in dark field electron microscopy. *Optik*, **38**, 335–50.

McGregor-Shaw, J. B., Easterbrook, K. B. & McBride, R. P. (1973). A bacterium with echinuliform (non prosthecate) appendages. *International Journal of Systematic Bacteriology*, **23**, 267–70.

Misell, D. L., Stroke, G. W. & Halioua, M. (1974). Coherent and incoherent imaging in the scanning transmission electron microscope. *Journal of Physics, Series D*, **7**, L113–L117.

Reimer, L. & Gentsch, P. (1975). Superposition of chromatic error and beam broadening in transmission electron microscopy of thick carbon and organic specimens. *Ultramicroscopy*, **1**, 1–5.

Rose, H. (1974). Phase contrast in scanning transmission electron microscopy. *Optik*, **39**, 416–36.

Rose, H. (1975). Zur theorie der Bildentstehung im Elektronenmicroskop. I. *Optik*, **42**, 217–44.

Schiske, P. (1968). Zur Frage der Bildrekonstruktion durch Fokusreihen. In *Proceedings of* 4*th European Regional Conference on Electron Microscopy, Rome 1968*, vol. 1, pp. 145–6. Tipografia Poliglotta Vaticana, Rome.

Someya, T., Goto, T., Harada, Y., Yamada, K., Koike, H., Kokubu, Y. & Wantanabe, M. (1974). On development of a 100 kV field emission electron microscope. *Optik*, **41**, 225–44.

Stroke, G. W. & Halioua, M. (1972). Attainment of diffraction-limited imaging in high-resolution electron microscopy by 'a posteriori' holographic image sharpening. I. *Optik*, **35**, 50–65.

Stroke, G. W. & Halioua, M. (1973). Image improvement in high-resolution electron microscopy with coherent illumination (low-contrast objects) using holographic image deblurring. *Optik*, **37**, 192–203, 249–64.

Stroke, G. W., Halioua, M., Thon, F. & Willasch, D. (1974). Image improvement in high-resolution electron micrscopy using holographic image deconvolution. *Optik*, **41**, 319–43.

Thomson, M. G. R. (1973). Resolution and contrast in the conventional and scanning high resolution transmission electron microscope. *Optik*, **39**, 15–38.

Thomson, M. G. R. (1975). Resolution and signal-to-noise ratio: a comparison between the conventional transmission electron microscope and the scanning transmission electron microscope. In *Physical Aspects of Electron Microscopy and Microbeam Analysis*, ed. B. M. Siegel & B. R. Beaman, pp. 29–46. Wiley, New York.

Zeitler, E. & Thomson, M. G. R. (1970). Scanning transmission electron microscopy. I and II. *Optik*, **31**, 258–80, 359–66.

R. P. FERRIER

Microanalysis of biological material using electron energy loss spectrometry

Elemental analysis of the thin specimens of interest in biological electron microscopy has in the past mainly been carried out by X-ray microanalysis. The basic physics underlying this method and its application to the study of biological materials are reviewed elsewhere in this volume by Chandler, Hutchinson and Appleton.

The sensitivity of X-ray microanalysis is, however, limited, particularly when the elements of interest are of low atomic number (Z). First, the fluorescence yield of the characteristic X-rays (ω), which may be defined as the number of X-ray quanta produced per ionisation of an electron in a particular atomic electron shell (the shells are labelled K, L, M, N, etc., in order of increasing radius), decreases rapidly with atomic number. The reason for this is that the emission of an X-ray is only one way in which the excess energy is released when an electron in an outer atomic shell falls into the hole in an inner shell created by the ionisation process. In the alternative mechanism the excess energy is given to another atomic electron which may then escape from the atom; these electrons are known as Auger electrons. Thus it is found that as Z decreases the process of Auger electron production becomes increasingly more probable to the extent that for sodium ($Z = 11$) 1 in 40 K-shell ionisations produce an X-ray and for carbon ($Z = 6$) the figure is only 1 in 400. The situation becomes even worse for other atomic shell excitations since $\omega_K > \omega_L > \omega_M$, etc. The second limitation is concerned with the X-ray collection efficiency which may be attained using currently available equipment. Thus for wavelength dispersive detectors the detection efficiency is only of the order of 10^{-3}–10^{-4}. For the more recently applied non-dispersive detectors the figure is probably more typically of the order of 10^{-2}, although Hutchinson (this volume) using a scanning electron microscope has achieved an efficiency $\sim 10^{-1}$. The third limitation which must be considered arises because, as the atomic number of an element decreases, the X-rays produced become more easily absorbed both by the specimen itself and by the windows which are present in most X-ray detectors. This phenomenon of absorption is one of the reasons why non-dispersive X-ray detectors are currently limited to the detection of elements of atomic number $\geqslant 10$.

Alternative methods are, however, available for microanalytical purposes. As discussed previously Auger electrons are produced within a specimen as a result of atomic excitation and their yield for low atomic number elements is high. Since the Auger electrons are characteristic of the elements, they can be, and indeed are, used for microanalysis. However the energies of the Auger electrons are low and the analyses are restricted to the examination of surface layers and therefore not strictly applicable to the types of specimen typically examined by electron microscopy. We shall not discuss Auger electron analysis further in this paper.

Another method which can be used for microanalysis is to study the energy lost by the electrons which have produced the ionisation events in a specimen, i.e. the events which result in the production of the X-rays and Auger electrons. The detection of those ionisation loss electrons offers a number of advantages when used for microanalysis, particularly when the elements of interest are mainly of low atomic number (e.g. beryllium, boron, sodium) and are distributed in a low atomic number matrix, as is the case for biological materials. First, the signal produced is not subject to absorption corrections due either to the specimen itself or to a detector window material, and the efficiency of collection can be very high ($\sim$ 0.1–1.0). In addition the 'fluorescence yield' is unity since each ionisation is the result of a particular electron undergoing the appropriate energy loss.

The purpose of this paper is to review the application of these ionisation loss electrons for microanalysis. Thus in the following section we consider aspects of the basic physics of the electron energy loss process with particular reference to the ionisation losses. This is followed by a description of the apparatus which is currently in use for such studies, together with estimates of the costs of setting up such equipment. Finally the possibilities of electron energy loss spectrometry for microanalysis of biological materials will be discussed and the probable sensitivities of this technique will be indicated.

Aspects of the physics of electron energy loss processes

Before we can discuss the application of electron spectrometry as a microanalytical tool we must have some appreciation of the electron energy loss process itself. The aspects which are of importance are the probability or cross-section of the particular type of event in which we are interested and the angular distribution of the inelastically scattered electrons which arise from the event as a function of such parameters as incident beam accelerating voltage and the specimen thickness and composition.

In the theory of scattering, the probability of a particular event is usually expressed in terms of a cross-section (σ) which has units of area per atom. Thus the probability for an inelastic scattering event in a specimen of thickness t is $Nt\sigma$ where N is the number of atoms per unit volume. When we wish to consider the factors which influence the use of energy loss electrons for microanalysis we require to know how the inelastic electrons are scattered as a function of the energy lost (E) and the angles of scattering (θ) relative to the incident beam direction. This may be expressed in terms of a cross-section per atom for scattering with an energy loss in the range $\mathrm{d}E$ about E and into a solid angle $\mathrm{d}\Omega$ about the direction θ. This is written as the doubly differential cross-section $\mathrm{d}^2\sigma/\mathrm{d}\Omega\,\mathrm{d}E$ and it can be shown that this quantity is given by (Hubbard, 1955)

$$\frac{\mathrm{d}^2\sigma}{\mathrm{d}\Omega \mathrm{d}E} = \frac{Kf[\epsilon(E,\theta)]}{E_0\,(\theta^2+\theta_{\mathrm{E}}^2)}, \qquad (1)$$

where K is a constant, E_0 is the incident electron beam energy and $\theta_{\mathrm{E}} = E/2E_0$ is the characteristic inelastic scattering angle. In equation (1) the energy dependence is governed by the energy loss function $f[\epsilon(E,\theta)]$ since the dielectric constant $\epsilon(E,\theta)$ contains all the information on the electronic properties of the material, and the momentum transfer is described by the factor $1/(\theta^2+\theta_{\mathrm{E}}^2)$.

Unfortunately due to a lack of knowledge of the precise form of the dielectric constant it is not generally possible to evaluate equation (1) directly and recourse must be made to approximations. There has been over the last sixty years considerable theoretical work concerned with the derivation of the scattering cross-sections for the ionisation losses which will be our particular interest and the reader is referred to Hills (1973) for a review of this aspect.

It should be stressed that equation (1) is applicable to *all* the energy loss processes occurring in the specimen. Of this totality of loss processes two distinct types have been utilised for microanalysis purposes, these being the losses associated with excitation of the valence electrons and those associated with ionisation of the elemental species. The spectra from these two types of process are illustrated in Fig. 1 where energy loss spectra from an iron–gadolinium alloy and from a thymine film ($C_5N_2O_2H_6$) are displayed. The structure in the energy loss region 0–40 eV is due principally to valence electron excitations. These appear either as discrete plasmon losses in substances in which there are a relatively large number of free valence electrons which behave in a collective manner, or as a combination of individual interband transitions and collective processes as in the transition elements, insulators, organic materials, etc. These losses are not found to be well lccalised

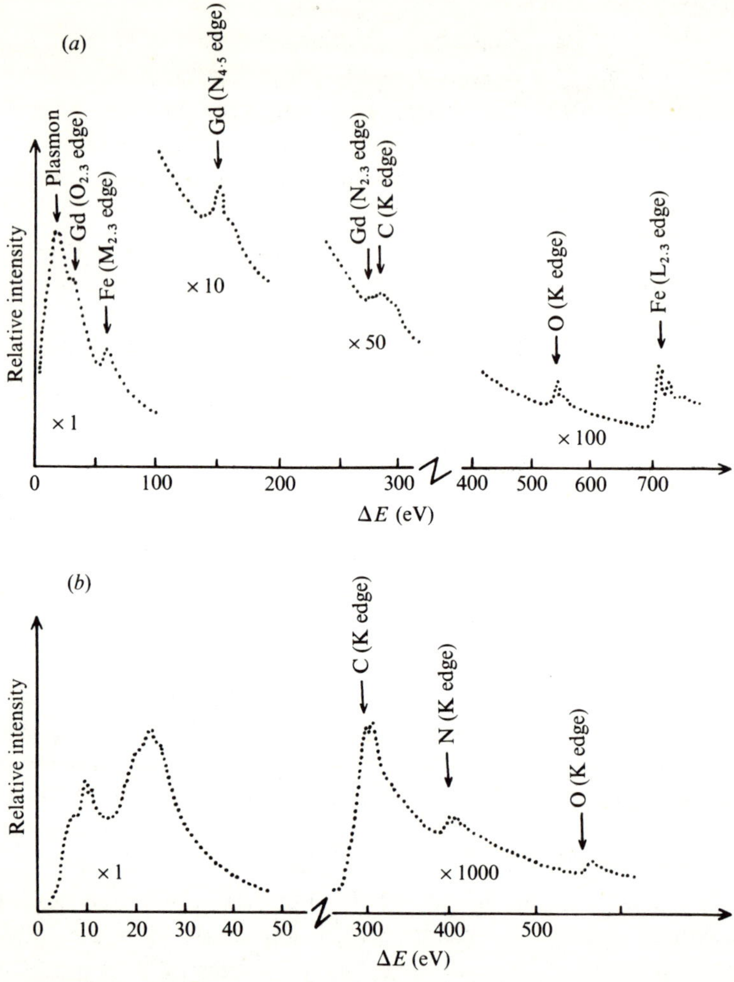

Fig. 1. Energy loss spectra obtained over a large energy range. They display contributions from the valence electrons and ionisation losses. (*a*) Iron–gadolinium alloy (from Colliex & Trebbia, 1976), (*b*) thymine (from Isaacson & Johnson, 1975).

nor are they in most cases definitively characteristic of the elements present. Their application for microanalysis purposes is therefore limited to special cases and these are principally in the metallurgical field (see for example Edington & Hibbert, 1973; Brousseau-Lahaye *et al.*, 1975). For biological materials their application is of secondary importance and we will not discuss these losses further in this paper.

The energy loss peaks shown in Fig. 1 which generally occur at higher

Table 1. *Comparison of theoretical and measured mean free paths for ionisation losses:* $E_0 = 60\ kV$

Element	Loss	Energy of loss (eV)	Theoretical mean free path (μm)	Experimental mean free path (μm)
Carbon	K	285	3.5	3.6 (±15 %)
Aluminium	K	1550	55	50 (±20 %)
Chromium	$L_{II,\ III}$	570	5.9	4 (±20 %)
Iron	$L_{II,\ III}$	710	7.1	6 (±30 %)
Copper	$L_{II,\ III}$	930	10.5	9 (±20 %)
Silver	$M_{IV,\ V}$	390	No data	9 (±30 %)

energies are those associated with ionisation of electrons in the various electron energy shells present in an atom. The energies at which these losses occur in the spectrum are characteristic of the elements present in the specimen. These ionisation losses are, for low atomic number elements, hardly deviated from the direction of incident electron beam and this is the reason for the high detection efficiencies which can be achieved. It is also relatively easy, particularly in the transmission electron microscope (TEM), to collect all the ionisation loss electrons and hence measure the total cross-section for any particular event. Such studies have been carried out by, among others, Leapman & Cosslett (1976) who measured the total cross-sections for various excitations in a range of elements. Their results are presented in Table 1 in the form of the mean free path (λ) for the particular event, the mean free path being related to the cross-section by

$$\lambda = \frac{1}{n\sigma} = \frac{A}{N_0\rho} \cdot \frac{1}{\sigma}, \tag{2}$$

where n, A, ρ, are respectively the number of atoms per unit volume, the atomic weight and the density of the specimen, N_0 is Avogadro's number and σ is the cross-section for the particular event. The important point to note is that the mean free paths are long in comparison to the thickness of the specimen (~ 50 nm) which would normally be examined in the electron microscope operated at the stated accelerating voltage. This implies that the number of electrons which lose energy in these events is a very small proportion of the total incident electron flux and hence we must ensure that the collection system is as efficient as possible. It should be noted, however, that the mean free paths for the equivalent characteristic X-ray emission will be even longer since the effect of fluorescence yield must be taken into account.

The angular distribution of the ionisation loss electrons has been the subject of a number of experimental studies (Wittry, Ferrier & Cosslett, 1969; Colliex & Jouffrey, 1972; Egerton & Whelan, 1974; Egerton, 1975). The results obtained indicate that the distribution of these electrons is more sharply peaked than the distribution of the 'background' loss electrons occurring in the same region of the energy loss spectrum. These background loss electrons are the result of non-characteristic energy loss processes occurring within the specimen, and a knowledge of their angular distribution is important since for a given region of the spectrum the ratio of the numbers of ionisation loss electrons to background electrons will be a function of the range of scattering angle accepted by the analyser which is used. It will be shown later that the minimum detectable mass fraction of a given element is proportional to the measured signal-to-noise ratio (S/N) which relates the intensity of the signal derived from ionisation loss electrons (I_P) to the intensity of the non-characteristic background loss electrons (I_B) measured over the same energy loss range of the spectrum. The signal-to-noise ratio has been defined in the literature in two ways, namely, $I_P/(I_B)^{\frac{1}{2}}$ or $I_P/(I_P+I_B)^{\frac{1}{2}}$. The values of I_P and I_B for a carbon film measured as a function of analyser angular aperture are shown plotted in Fig. 2 (Egerton & Whelan, 1974) and the results for S/N as a function of analyser angular aperture are displayed in Fig. 3 for the elements boron, carbon and oxygen. It should be noted that the angle at which the maximum signal-to-noise ratio occurs increases with the energy of the ionisation loss detected, in agreement with the $1/(\theta^2 + \theta_E^2)$ dependence of equation (1). It should be stressed that we have been talking about detectability of an element rather than quantitative microanalysis. If we are to be concerned with the latter we must know the proportion of the ionisation electrons which we have collected with our optimum analyser aperture or find some other means of putting our results on an absolute basis. This must be the subject of a great deal more research.

Another parameter which must be considered in relation to the quality of the ionisation loss signal is the specimen thickness. The influence of thickness may be illustrated with reference to Fig. 4 which shows energy loss spectra obtained from single, double and triple thicknesses of a carbon film (~ 20 nm single thickness). These signals were obtained (Ferrier & Hills, 1975) using a constant incident beam intensity for all three spectra. Two main effects may be noted: the first is that the signal-to-background ratio (and hence S/N ratio) decreases as the specimen thickness increases; the second effect is that due to plural scattering the definition of the loss peak is reduced. These two effects are observed in the majority of specimens, particularly when the thickness approaches that of the plasmon mean free path at the accelerating voltage used (~ 50 nm) and hence the specimen thickness should be limited to a value less than this for optimum element detectability.

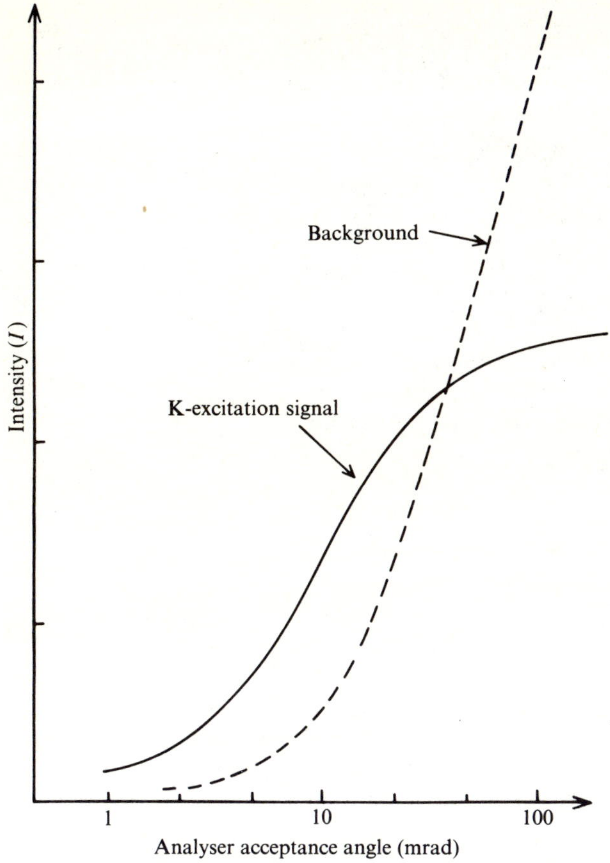

Fig. 2. The intensity of K-ionisation loss and the background measured as a function of the acceptance angle of the energy analyser for a thin carbon film. Incident electron energy 80 keV. (After Egerton & Whelan, 1974.)

The influence of the inelastic scattering processes on the use of ionisation loss electrons for microanalysis may be summarised as follows:

(i) The cross-sections for the production of ionisation loss electrons are small and hence we must ensure that our detection system is optimised.

(ii) The distribution of the ionisation loss electrons is more sharply peaked in the forward direction than that of the background loss electrons. This implies that the efficiency of detection can be high, but we must limit the angular acceptance of the analyser to achieve an optimum S/N ratio.

(iii) When the specimen thickness is increased the S/N ratio decreases

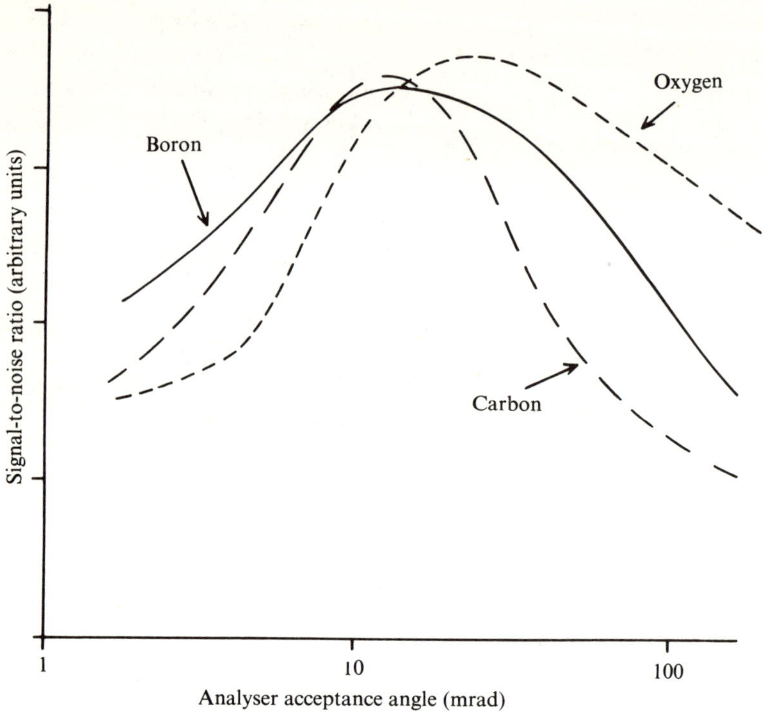

Fig. 3. The signal-to-noise ratio, $I_P/(I_P+I_B)^{\frac{1}{2}}$, obtained for boron, carbon and oxygen as a function of the analyser acceptance angle. The analyser was set to accept energy losses over a range 35 eV from the inception of the K-loss peak (After Egerton, Roussouw & Whelan, 1976.)

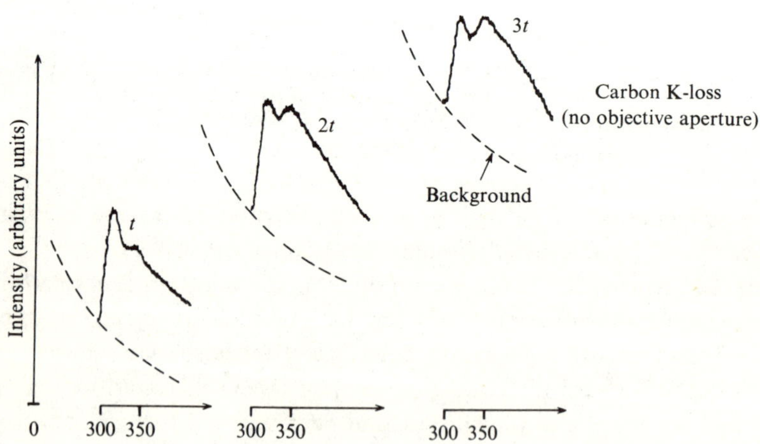

Fig. 4. The influence of specimen thickness (t) on the signal-to-background ratio for the carbon K-loss. Incident electron energy = 60 keV. (After Ferrier & Hills, 1975.)

due to the influence of the background, which increases more rapidly than the signal, and the redistribution of intensity in the ionisation loss peak due to plural scattering. Hence for optimum element detectability we require thin specimens of thickness less than the mean free path for plasmon excitation.

Experimental arrangements

In the TEM two main approaches have been followed to allow energy loss analysis utilising the ionisation loss electrons. In one of these the energy analyser is located after the objective lens but before the projector lenses. This type of analyser was originally introduced by Castaing & Henry (1962) and has also been used by Colliex & Jouffrey (1972) and by Egerton & Whelan (1974).

The principle of the analyser is illustrated in Fig. 5. The electron beam which exits from the cross-over below an intermediate lens enters the first half of a triangular magnetic prism in which the induction is uniform. After the first 90° deflection the beam exits normally to the vertical prism face. A reflection at an electrostatic mirror brings the beam back into the prism where a second deflection through 90° occurs and the beam is brought back on axis. The result of this trajectory is that an energy loss spectrum is formed at the level of the exit cross-over from the prism and a particular energy band may be selected to form the image by means of an aperture placed in this plane.

The instrument discussed above is essentially an energy selecting electron microscope. An alternative approach is the energy analysing electron microscope which is a simpler device in which we locate an energy analyser below the viewing screen of a conventional TEM. This is the approach adopted by the Cambridge group (Wittry, 1969) and by some of the commercial manufacturers, as will be discussed later. The essential basis of the system (for a detailed discussion see Hills, 1973) is shown in Fig. 6. The energy analyser is once again a magnetic prism wth a maximum energy resolution of 5 parts in 10^5, which is adequate for analysis of the ionisation losses. The pole faces of the prism are shaped to give two-dimensional focusing and the cross-over below the projector lens is imaged in the plane of the exit aperture whose size determines the energy resolution of the spectrum. An aperture below the viewing screen selects the area of the specimen to be analysed and this may be varied by varying the image magnification selected. The angular acceptance of the analyser may be varied by selecting objective apertures of different sizes. Although in this case it would be possible to sweep the energy loss spectrum over the detector

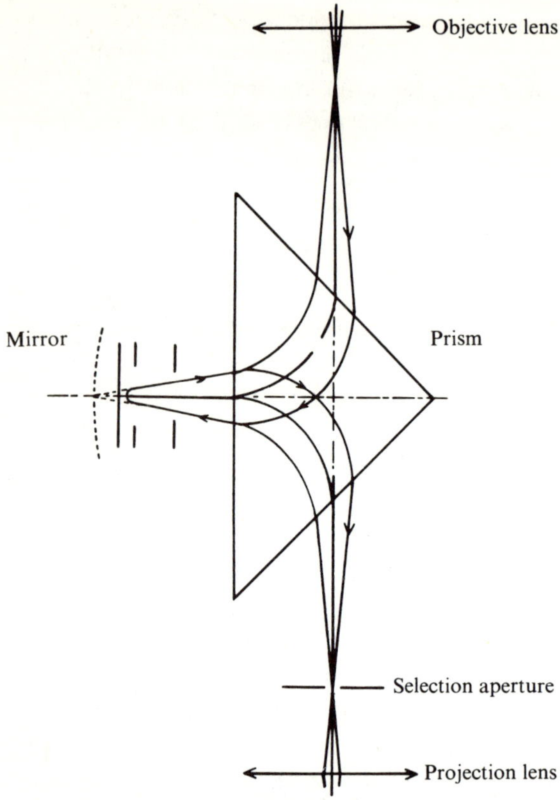

Fig. 5. The energy analyser of the energy selecting electron microscope. (After Castaing & Henry, 1962.)

aperture by varying the prism excitation, it is sometimes preferred to maintain the prism excitation constant at a value appropriate to the detection of the zero loss electrons. The accelerating voltage is then swept while maintaining all imaging lens supplies constant. In this way the influence of chromatic aberration is considerably reduced since the lenses are always 'in focus' for the particular energy loss electrons being detected.

It should be stressed that in any system used to detect ionisation losses it is essential that the detection system (normally a scintillator–photomultiplier combination) is optimised, and to achieve this great care must be exercised in its design. This is also the reason for preferring a magnetic prism to an analyser tied to the accelerating voltage, since this avoids problems of micro-discharge noise associated with energy analysers which are tied to the electron accelerating voltage.

In recent years the scanning transmission electron microscope (STEM) has become of increasing importance, particularly when it is fitted with a

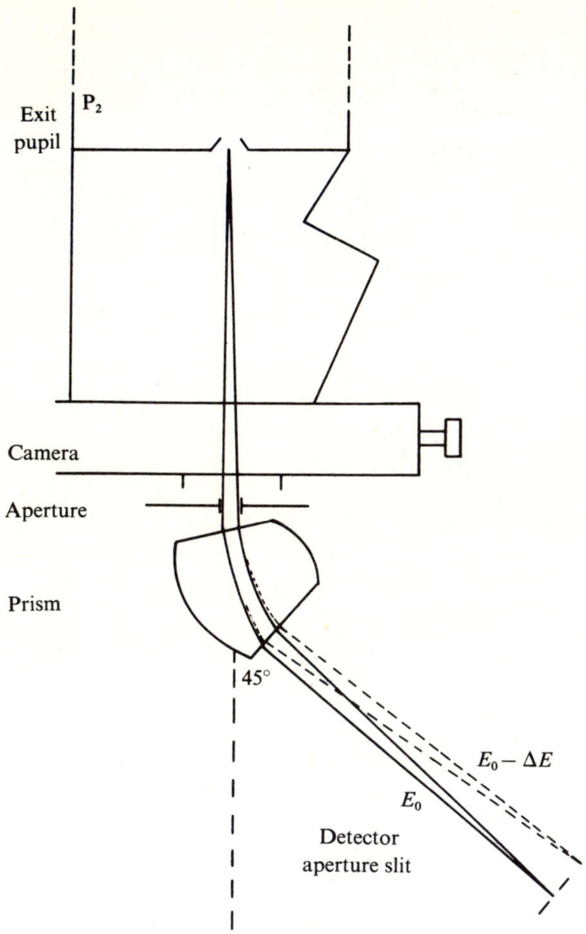

Fig. 6. The prism spectrometer attachment. (After Wittry, 1969.)

field emission gun (Crewe, Isaacson & Johnson, 1971). This system holds considerable promise for microanalysis using ionisation loss electrons, as will be discussed later. In such a system (Fig. 7) there is only one place where we can put the energy analyser and that is below the probe-forming lens. One restriction of this arrangement is that there will be a maximum size of angular aperture appropriate to a given acceptable level of analyser aberrations. Thus, unlike the energy analysing TEM, there is only limited scope for varying the analyser acceptance angle unless an additional lens is provided after the probe-forming lens.

Commercial versions of the STEM with field emission guns operating at voltages of up to 100 kV are now available and all are fitted with energy

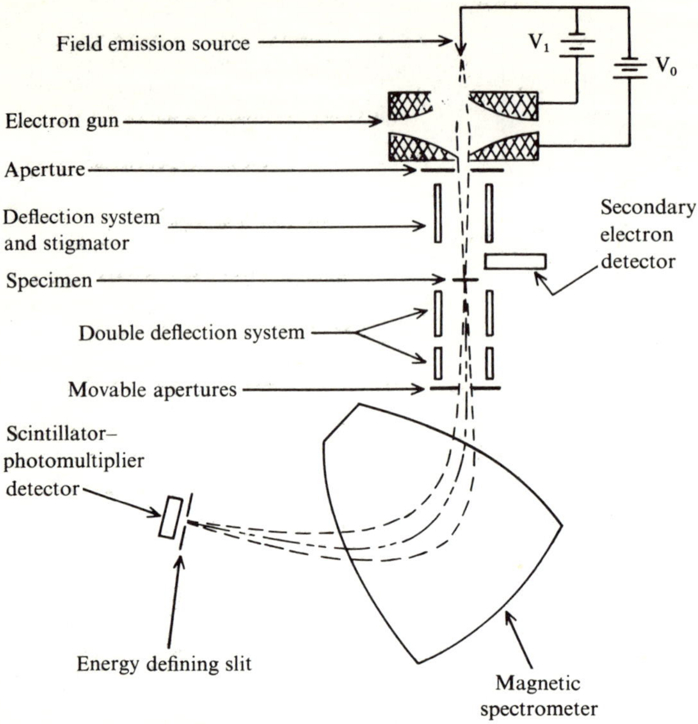

Fig. 7. The scanning transmission electron microscope with energy analyser. (After Crewe *et al.*, 1971.)

analysers (see for example the illustration of VG Instruments' system in the article by Elder in this volume). However, these instruments which require the provision of an ultra-high vacuum are extremely expensive, the price being in the range £100000–£200000. To the author's knowledge no commercial version of the Castaing–Henry system is available, but several manufacturers of high resolution TEMs do now market analysers which can be fitted below the viewing screen and these are suitable for the type of analysis discussed here. The total price for such an accessory including the associated counting and display systems is believed to be in the region of £15000–£30000.

The field of electron energy loss spectrometry, and in particular its commercial exploitation, is likely to change rapidly over the next 12–18 months, and hence it was decided not to give a list of equipment currently available.

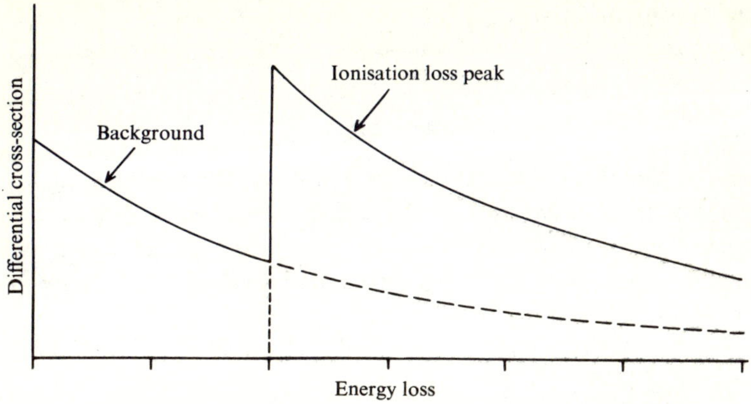

Fig. 8. Schematic illustration of ionisation loss peak. The background in the same energy range in the spectrum is shown as the dashed line.

Microanalytical aspects of electron spectrometry

The application of electron spectrometry as a microanalytical tool, particularly in the study of biological specimens, may be discussed with reference to Fig. 8. This shows a schematic representation of an ionisation loss peak from a sample superimposed on a background. What we wish to investigate is the smallest signal which we can detect in the presence of the background. Our main interest will be for the case where the signal arises from an element, e.g. sodium, present in a biological matrix which may be taken to be carbon. However, let us first consider the sample as a pure element and calculate the smallest number of atoms of that element which can be detected; we may denote this, following Isaacson & Johnson (1975), by the minimum detectable mass or MDM. Alternatively we may consider the more important case of an element present as one component in a matrix of other elements. In this case we are interested in the minimum mass fraction (MMF) which can be detected and the major difficulty in estimating this parameter lies in evaluating the background contribution which will arise mainly from the matrix, particularly if the element is present in small quantities.

Let us first discuss minimum detectable mass. To calculate this quantity exactly we would have to take into account the background present in the spectrum in the neighbourhood of the ionisation loss peak. However to a reasonable approximation we may neglect the effect of background and merely calculate the smallest mass of an element which will give rise to a signal which may be detected with a given total measurement time, the latter being consistent with such factors as instrument stability, specimen radiation damage, etc.

The maximum current (i_s) into the analyser due to K loss electrons is, assuming that the specimen thickness $t \ll \lambda_K$ the mean free path,

$$i_s = J\frac{\pi d^2}{4} \cdot \frac{t}{\lambda_K}, \tag{3}$$

where J is the electron beam current density at the specimen and d is the diameter of the specimen area contributing to the signal. If we assume that this specimen area is defined by an electron probe of diameter d and semi-angle of illumination α, then the probe current (I) is related to the electron gun brightness β by

$$\beta = \frac{4I}{\pi^2 d^2 \alpha^2}. \tag{4}$$

It then follows from equations (2), (3) and (4) that the number of atoms n_A giving rise to a signal of n electrons per second is

$$n_A = \frac{ne}{\beta\pi\alpha^2\sigma} = \frac{ne}{J\sigma_K}, \tag{5}$$

where e is the electronic charge. Table 2 gives the calculated values of MDM for two values of current density, one appropriate to the brightness of the conventional thermionic electron gun ($\beta = 10^9$ A m^{-2} steradian^{-1}) fitted to all but a few microscopes and one consistent with the brightness of a field emission source ($\beta = 10^{13}$ A m^{-2} steradian^{-1}). It should be noted that these calculated values represent the ultimate detectability and in practice the values which can be attained are limited by a number of factors which include:

(1) The background which arises mainly from interband transitions must be taken into account. If this is done the MDM will increase.

(2) The proportion of the ionisation loss electrons which are collected will depend on the angular aperture of the energy analyser used and this will in general decrease with atomic number. In calculating the proportion collected one should also take into account the semi-angle α of the illumination cone forming the probe on the specimen. For a given brightness the probe current density can only be increased by increasing α and this can reduce the proportion of the ionisation loss electrons collected.

(3) The signal from the ionisation loss electrons is found to extend over an energy range $\sim$ 200 eV. Ideally, to achieve small values for the MDM we would wish to record all these electrons simultaneously while retaining a reasonable energy resolution ($\sim$ 5 eV). This would require a parallel recording system. One way to achieve this would be to record the spectrum photographically, but a more desirable solution

Table 2. *The minimum detectable number of atoms, n_A, and the minimum detectable mass (MDM)*

	Atomic number	5	6	12	13
	σ_K (10^{-24} m^2)	3.9	2.5	0.38	0.30
$J = 5.6 \times 10^3$ A m^{-2}	n_A	730	1100	7500	9500
	MDM (10^{-20} g)	1.3	2.3	31	43
$J = 10^6$ A m^{-2}	n_A	4	6	42	53
	MDM (10^{-22} g)	0.73	1.3	17.3	24.0

Calculated from equation (5) on the basis of $n = 100$ electrons s^{-1} and a total recording time of 10 s giving an accuracy of 3 %. Two values of current density are used $J = 5.6 \times 10^3$ A m^{-2} for a thermionic emission gun and $J = 10^6$ A m^{-2} for a field emission gun. $E_0 = 60$ kV

would be to accomplish this electronically and this is currently under investigation.

Isaacson & Johnson (1975) have carried out calculations which attempt to take into account some of the points discussed above. Their results are shown in Fig. 9, which also shows the MDM which can be achieved using X-ray microanalysis. It should be noted that their calculated values are still somewhat optimistic and will certainly not be reached until a good recording system is available.

To calculate the minimum mass fraction (MMF) of an element in a biological matrix we require to estimate the background in the neighbourhood of the loss peak. Isaacson & Johnson have carried out such a calculation for a range of elements assuming that the matrix is carbon. Their conclusion is that the background arises principally from the high-energy tail of the characteristic K ionisation peak (at 280 eV) of the carbon matrix. They set a criterion for detectability that the signal-to-noise ratio (defined as $I_P/(I_P)^{\frac{1}{2}}$) must not be less than 2. Their analysis indicates that the MMF is proportional to the current in the probe rather than the current density, as was the case for the MDM, and their results for a beam energy of 25 keV and two values of probe current 10^{-6} A and 10^{-8} A are shown in Fig. 10. It should be noted that with a field emission source these currents could be obtained in probes of sizes of ~ 100 nm and ~ 10 nm respectively, assuming a semi-angle of illumination of 0.01 rad. Thus for the detection of small quantities of material in highly localised regions of the specimen, the field emission gun is essential. Isaacson & Johnson have also carried out experiments to try to confirm these theoretical predictions. They have measured the S/N ratio for a number of elements in specimens of known composition and then estimated the mass fraction of any particular element which would have given a S/N ratio of 2.

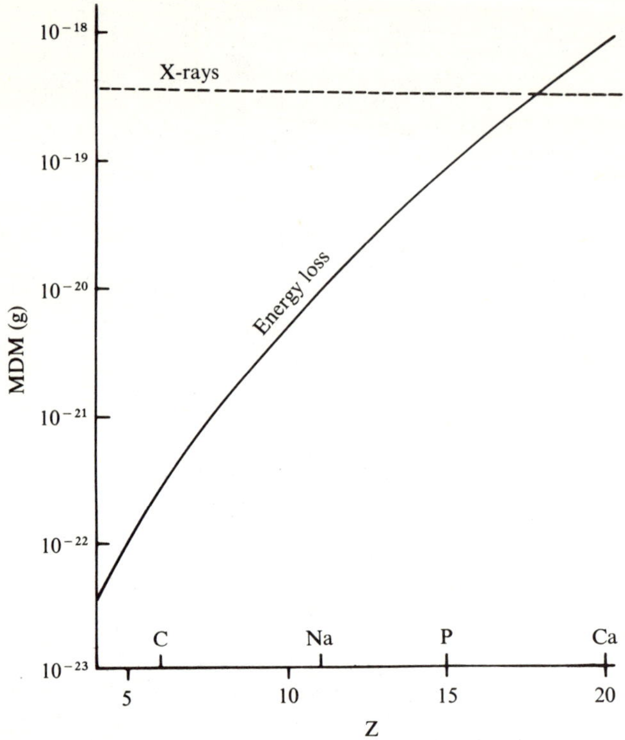

Fig. 9. The minimum detectable mass (MDM) as a function of atomic number (z) for 25 keV electrons. The curve labelled 'energy loss' was calculated on the basis of the following assumptions: (i) the semi-angle of acceptance of the energy analyser was 0.025 rad; (ii) the energy window of the analyser was variable and equal to $\frac{1}{2}[E_K(Z)-E_K(Z-1)]$ where E_K is the K-ionisation energy; (iii) current density $J = 10^6$ A m^{-2}; counting rate $=$ 100 events s^{-1}. The curve labelled 'X-rays' was calculated assuming the use of a 100 % efficient non-dispersive X-ray detector subtending an angle of $\frac{1}{4}\pi$ rad at the specimen. (After Issaacon & Johnson 1975.)

These results indicate that their predictions were overoptimistic by a factor of $\sim$ 3. Even allowing a more conservative factor this still implies that the number of biological systems which can be investigated is quite considerable and certainly much greater than is accessible by X-ray microanalysis.

It should be stressed once again that what we have been discussing is the detectability of elements in particular circumstances, and we have not made any predictions on the accuracy which could be achieved in quantitative microanalysis; indeed at this particular time it is not possible to do so with any real confidence, although within the next few years this will undoubtedly become feasible.

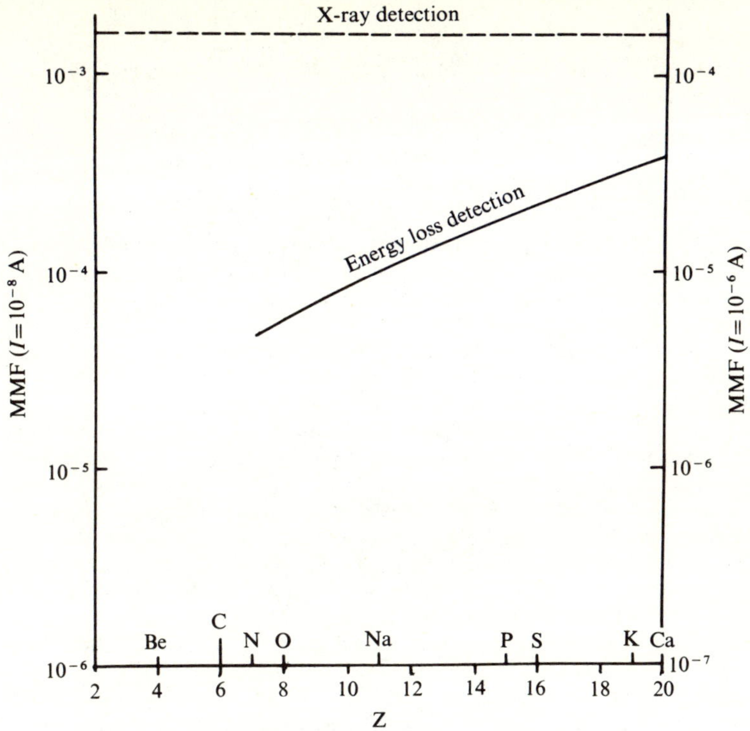

Fig. 10. The minimum mass fraction (MMF) of a given element which can be detected in a carbon matrix using 25 keV electrons. The left-hand vertical scale corresponds to probe current $I = 10^{-8}$ A and the right-hand scale to $I = 10^{-6}$ A. The calculation assumes an analyser semi-angle of acceptance of 0.025 rad, a counting time of 100 s, S/N = 2, energy window of analyser = $E_K/10$. The X-ray curve was calculated on the same assumptions as those applying to Fig. 9.

Summary

It has been shown that the technique of electron spectrometry utilising the elemental ionisation energy losses promises the provision of a powerful microanalytical tool for the study of biological systems in which the elements of interest are of low to medium atomic number. The range of systems which can be studied is large, particularly if a microscope fitted with a field emission gun is available. However the cost of these will prohibit their widespread use and therefore progress at the ultimate levels of element detectability is likely to be slow. Whatever microscope is used for these studies the performance would be considerably enhanced if parallel electronic detection systems were to become available.

The author wishes to thank J. N. Chapman, V. E. Cosslett and R. P. Leapman for helpful discussions on various aspects covered by this review.

References

Brousseau-Lahaye, B., Colliex, C., Frandon, J., Gasgnier, M. & Trebbia, P. (1975). Determination of the electron excitation spectrum in scandium and yttrium by means of characteristic energy loss measurements. *Physica Status Solidi*, **69B**, 257–66.

Castaing, R. (1971). Secondary ion microanalysis and energy-selecting electron microscopy. In *Electron Microscopy in Material Science*, ed. U. Valdre, pp. 102–61. Academic Press, New York & London.

Castaing, R. & Henry, L. (1962). Filtrage magnétique des vitesses en microscopic electronique. *Comptes rendus hebdomadaires des séances de l'Academie des sciences*, **255**, 76–8.

Colliex, C. & Jouffrey, B. (1972). Diffusion inelastique des electrons dans un solide par excitation de niveaux atomique profonds. I. Spectre de pertes d'energie. *Philosophical Magazine*, **25**, 491–511.

Colliex, C. & Trebbia, P. (1976). Energy loss spectroscopy in the electron microscope as tool for microanalysis. In *Developments in Electron Microscopy and Analysis*, ed. J. A. Venables, pp. 123–8. Academic Press, New York & London.

Crewe, A. V., Isaacson, M. & Johnson, D. (1971). A high resolution electron spectrometer for use in the transmission scanning electron microscopy. *Review of Scientific Instruments*, **42**, 411–20.

Edington, J. W. & Hibbert, H. (1973). High resolution microanalysis of aluminium solid solution alloys using combined electron microscopy and energy analysis. *Journal of Microscopy* **99**, 125–46.

Egerton, R. F. (1975). Inelastic scattering of 80 keV electrons in amorphous carbon. *Philosophical Magazine*, **31**, 199–215.

Egerton, R. F., Roussouw, C. J. & Whelan, M. J. (1976). Progress towards a method for the quantitative microanalysis of light elements by electron energy-loss spectrometry. In *Developments in Electron Microscopy and Analysis*, ed. J. A. Venables, pp. 129–132. Academic Press, New York & London.

Egerton, R. F. & Whelan, M. J. (1974). High resolution microanalysis of light elements by electron energy loss spectrometry. In *Electron Microscopy 1974*, vol. 1, pp. 384–5. Australian Academy of Science, Canberra.

Ferrier, R. P. & Hills, R. P. T. (1975). Selected area electron spectroscopy. In *Advances in Analysis of Microstructural Features by Electron Beam Techniques*, pp. 41–66. Institute of Metals, London.

Hills, R. P. T. (1973). Electron spectrometry in the electron microscope. PhD Thesis, University of Cambridge, England.

Hubbard, J. (1955). The dielectric theory of electronic interactions in solids. *Proceedings of the Physical Society of London*, **68**, 976–86.

Isaacson, M. & Johnson, D. (1975). The microanalysis of light elements using transmitted energy loss electrons. *Ultramicroscopy*, **1**, 33–52.

Leapman, R. D. & Cosslett, V. E. (1976). Electron energy loss spectrometry; mean free paths for some characteristic X-ray excitations. *Philosophical Magazine*, **33**, 1–10.

Wittry, D. B. (1969). An electron spectrometer for use with the transmission electron microscope. *Journal of Physics, Series D*, **2**, 1957–66.

Wittry, D. B., Ferrier, R. P. & Cosslett, V. E. (1969). Selected area electron spectrometry in the transmission electron microscope. *Journal of Physics, Series D*, **2**, 1767–73.

T. E. HUTCHINSON

Energy dispersive X-ray microanalysis

This paper will be concerned chiefly with elemental microanalysis of frozen hydrated biological tissue (see Appleton, this volume, for a discussion of cryo-ultramicrotomy) but will apply equally well to the analysis of freeze-dried and fixed tissue, although implementation in each of these systems has its own particular problems (for discussion of these, see Chandler, this volume).

Existing techniques

Energy dispersive X-ray detector

The key component in any energy selective X-ray detection system is a lithium drifted silicon crystal. X-rays falling upon this crystal give rise to electron-hole pairs which are collected as a current pulse, amplified and transmitted to a pulse height analyser. The number of electron-hole pairs generated as a result of impingement of the X-ray is determined by the X-ray energy. The energy of the X-ray released by an atom excited by, for instance, electron irradiation, is dependent upon the elemental atomic number. Thus the number of X-rays of a particular energy incident upon the crystal per unit time is a measure of the concentration of the element giving rise to X-rays of that characteristic energy within the volume irradiated by the exciting electrons. This is displayed on a cathode ray tube as a spectrum of X-ray energy versus the integral number of such X-rays impinging on the crystal. Since the energy selective crystal does not depend upon Bragg diffraction of X-rays incident upon it at a mechanically set angle, as does the wavelength dispersive collection system (see Chandler, this volume), a full spectrum of elemental analysis can be obtained of a wide range of elements simultaneously This is especially important in relation to electron beam irradiation damage, referred to in the section on elemental loss during analysis (p. 221).

Several parameters of great significance to the choice of an X-ray detector crystal for analysis are given below.

(1) The number of X-rays detected by the crystal is determined by the solid angle subtended by the crystal from the X-ray source. This is clearly determined by both the distance of the detector from the source and the detector area. In order to obtain the greatest solid angle, the distance between the detector and source should be minimised and the area maximised. Currently, 80 mm^2 crystals are available. However, 30 mm^2 crystals are much preferred, since it is frequently possible to situate the crystal much closer to the specimen.

(2) The resolution in energy of the crystal is an important parameter. Greater resolution is possible with smaller crystal diameters. Resolutions of roughly 150 eV are commercially available. This is certainly adequate for most analyses, since characteristic energies of biologically important elements differ by energies greater than this. However, the primary energy peak of calcium overlaps with the energy of a secondary peak of potassium to a degree less than the resolution of the present detector systems, thus making quantification of small concentrations of calcium in the presence of potassium difficult.

(3) The X-ray detector crystal must be protected from the environment of the electron microscope by a thin beryllium window in order to prevent damage due to contamination. Elements of low atomic number, i.e. less than sodium, yield characteristic X-rays of very low energy. This energy is sufficiently low to be absorbed in the window and thus the detector efficiency for these elements drops to zero. Clearly a minimum thickness of detector window should be chosen to maximise the efficiency of collection for low atomic number species. However, excitation of the silicon crystal by scattered electrons remains a problem with very thin windows and 'nude' detectors yielding high Bremsstrahlung* backgrounds should be avoided.

A number of manufacturers supply energy dispersive X-ray microanalytical equipment and this intrumentation can be interfaced with most commercial makes of electron microscope. The current cost varies from approximately £12000 to £20000 or more, depending largely upon data processing accessories purchased.

Electron microscopes

Two generic types of electron microscopes are presently employed in elemental microanalysis. These are the transmission electron microscope (TEM)

* Bremsstrahlung arises from the deceleration of electrons in the presence of atoms and gives rise to non-characteristic X-radiation (see the chapter by Chandler, this volume).

equipped for operation with a small spot, i.e. roughly 3 nm, and the scanning electron microscope (SEM) having a spot size somewhat larger. SEMs offer the advantage of a large volume surrounding the specimen, available for cold stage and X-ray detector systems, due to the large distance (termed 'working distance') between the final lens and the specimen. This allows a much closer proximity of the X-ray detector to the specimen. They offer the further advantage that the scanning transmission capability is a standard component of the equipment, whereas this capability in TEMs is an added and rather expensive accessory. The major advantages of the transmission instrument over the scanning instrument are that electron diffraction capabilities used to determine the state of tissue hydration are standard and that a much smaller electron beam size is available. Scanning transmission imaging at a spot diameter of roughly 2 nm has now been attained and is clearly of significant advantage in determination of position for elemental analysis of ultrafine structures.

Frozen hydrated sectioning and transfer to the electron microscope

A major advance has recently been attained in the determination of elemental concentrations on a scale small with respect to the size of a cell by the techniques of rapid freezing, sectioning of the frozen hydrated tissue, and transfer to and observation of that tissue in the electron microscope in the frozen state. Frozen hydrated sectioning was first pioneered by Christensen (1971) and has remained virtually unchanged apart from minor refinements since his original work. Major advances in the techniques of transfer of the frozen hydrated sections to the electron microscope and in the development of cold stages for the electron microscope have been carried out in Cambridge (Echlin & Moreton, 1974) as well as other laboratories including our own (Hutchinson, Bacaner, Broadhurst & Lilley, 1974). The scheme adopted for transfer in our laboratory is shown in Fig. 1. The frozen specimen, mounted on an electron microscope grid, is fitted into a small specimen holder. The specimen holder is then withdrawn into a long tube of high heat capacity, pre-cooled to the temperature of liquid nitrogen, and transferred to the cold stage of the electron microscope. The tube thus maintains the low temperature of the specimen to a large degree and in addition provides a trap preventing moist ambient air from reaching the specimen. The specimen is inserted subsequently into the cold stage of the instrument, exposed only to a dry, cold ambient atmosphere. This scheme provides a simple and effective means of transfer of frozen hydrated material.

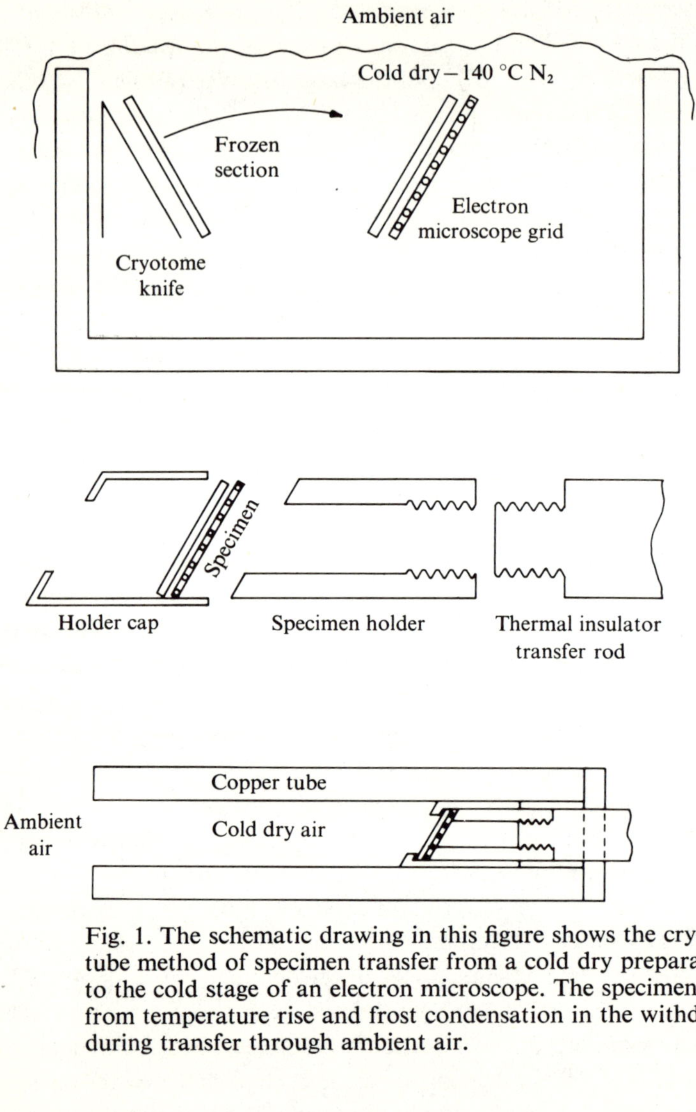

Fig. 1. The schematic drawing in this figure shows the cryogenic tube method of specimen transfer from a cold dry preparation area to the cold stage of an electron microscope. The specimen is protected from temperature rise and frost condensation in the withdrawn position during transfer through ambient air.

Structural localisation by imaging

In order to determine the location of structural elements to be analysed for elemental concentrations it is necessary to form an electron image of the frozen section. An example of such imaging is shown in Fig. 2 of frozen hydrated bovine skeletal muscle of thickness roughly 100 nm. It is likely that some degree of dehydration occurred prior to imaging of the specimen. Virtually all of the expected microstructural features are clearly imaged by

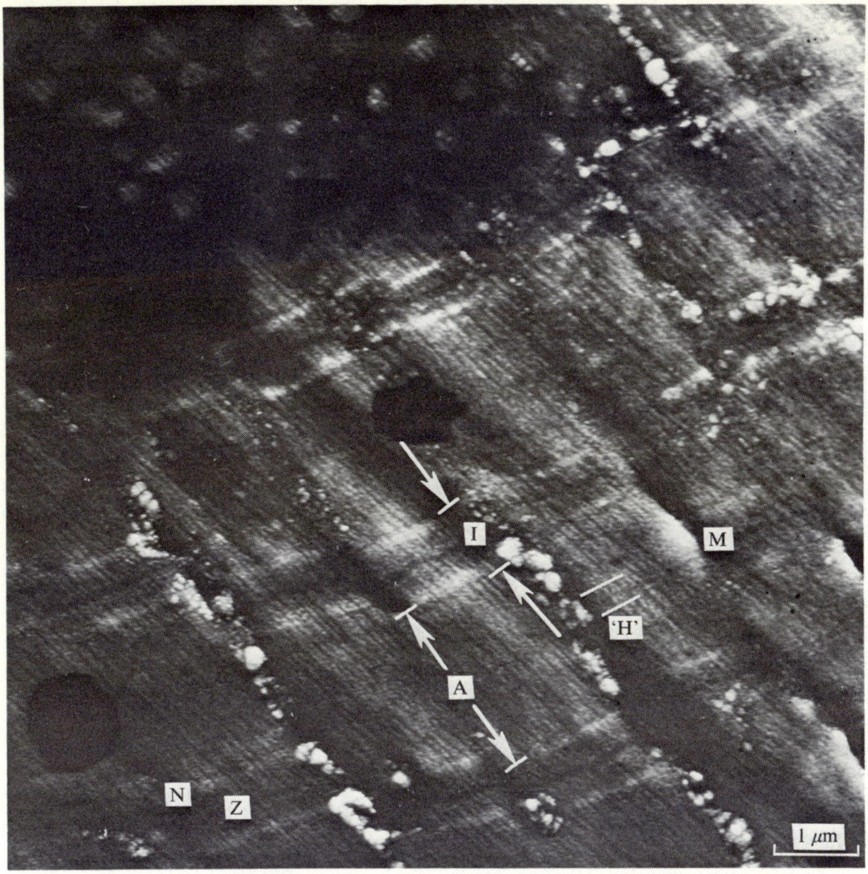

Fig. 2. The micrograph shown in this figure was recorded from a 100-nm thick frozen hydrated section of bovine skeletal muscle. Virtually all structural features normally observed in conventionally prepared muscle are present. A, A-band; I, I-band; H, H-band; M, M-line; N, N-line; Z, Z-line.

this technique, with resolution bordering on that obtained by conventionally fixed, embedded and stained material. An exception is the thick and thin filament system between the z-lines. The structures noted in this micrograph are likely to be the result of ice crystal formation during freezing but on a scale sufficiently small so as not to change ion positions appreciably. The contrast arises from local variations in electron scattering from proteins and frozen fluids and is thus low except where concentrations of heavier elements may exist, as in the Z-line of muscle. Partial dehydration gives rise to higher contrast since variations in the degree of binding of fluids, particularly water, exist within the structure leading to preferential sublimation and thus local

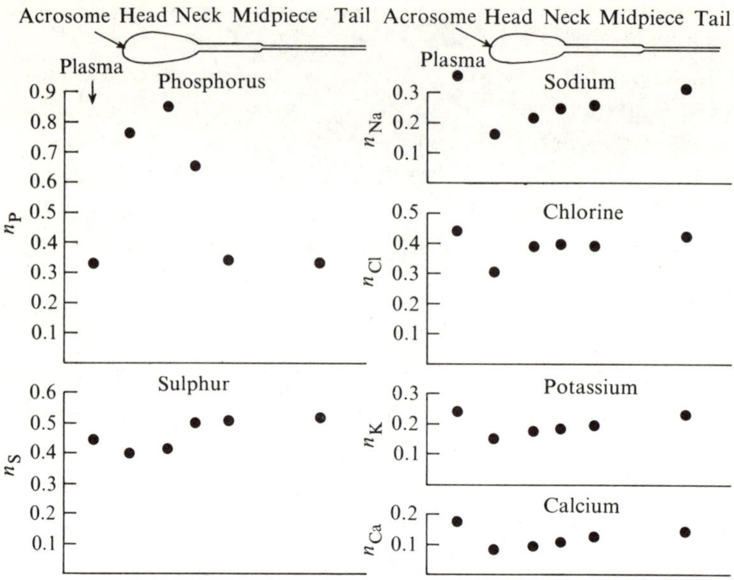

Fig. 3. Phosphorus, sulphur, sodium, chlorine, potassium and calcium in neat bovine spermatozoa were analysed with an energy selective X-ray analyser (Kevex). The sperm were frozen between two Formvar films on a beryllium grid in slush nitrogen 2 hours after ejaculation and examined at -150 °C with beam current of 1.3×10^{-9} A for approximately 100 s. The average n_X of eight analyses was calculated where $n_X = (P-b)_X$/Bremsstrahlung and Bremsstrahlung is the number of counts between 5.0 and 6.0 keV. P – b, peak minus background; X, the element under consideration.

differences in net mass scattering thickness. Recent studies in dark-field imaging of frozen hydrated sections indicate that significant dehydration is not necessary, however, for structural localisation (T. E. Hutchinson, D. Johnson and H. Ris, unpublished data).

Concentration data: acquisition and quantification

In order to determine concentrations of elements within the tissue it is first necessary to locate the electron probe, of diameter 10 nm, in a particular volume of the specimen. X-rays generated by electron beam interaction are collected by the X-ray detector and displayed as a spectrum, as shown in Fig. 3, of X-ray energy versus number of X-rays received by the detector. These particular data were collected from frozen hydrated bull sperm and are representative of the concentration of elements within preselected points within the sperm. The elements of biological interest, including sodium and elements having higher atomic numbers, are present. The relative concentrations of each of these elements are easily determined by measurement of the area of the peak and comparison with previously prepared standards. These standards are prepared by a technique of rapidly freezing aqueous solutions of known concentration between two thin collodion films. The areas of the peaks for known concentrations of elements (in the millimolar range) can thus be determined and directly compared to data obtained from biological unknowns. The sperm specimens are also obtained by a similar technique, which has been applied to numerous other single cell and culture cell preparations.

Limits of detection

The limits of detection in concentration of elements having an atomic number higher than sodium have been determined in our laboratory (J. Borek & T. E. Hutchinson, in preparation). These data, shown in Fig. 4, were obtained through implementation of the standard solution technique outlined in the previous section, by reducing the concentration to such a level that the peak could no longer be separated from the background. The limit of detections are clearly dependent upon the atomic number of the element present but can be roughly determined to be of the order of 1.5 mM for elements with a higher atomic number than sodium and 3–5 mM for sodium. Since numerous biological problems can be solved with these concentration limits it clearly makes the technique of tremendous importance to biological investigations. On the other hand, numerous experiments of great interest are outside the scope of present techniques solely because of this detection limit constraint, and advanced systems having lower detection limits are clearly needed.

Major advantages

Volume analysed

No other technique can compete with electron microscope probe analysis in respect of the smallness of the volume analysed. The technique is limited primarily by the electron scatter in the specimen. The diameter of the emergent electron beam from the bottom of the specimen is of the order of 0.5 times the thickness of the specimen. The volume which is subject to electron excitation is a right circular cone with height equal to the specimen thickness and base equal to 0.5 of the thickness, which for isolated cell membranes and very thin sections can be of the order of 10 nm.

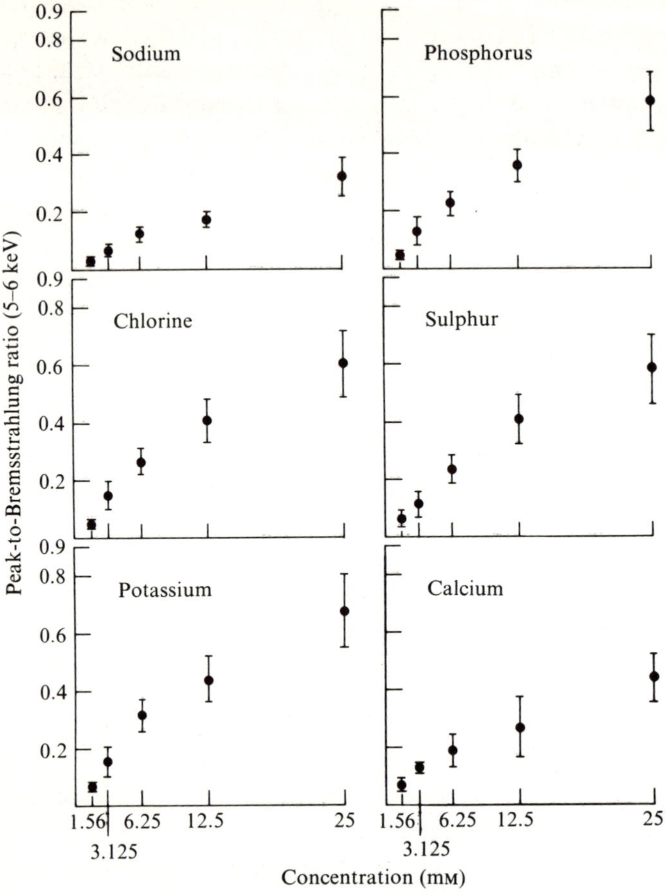

Fig. 4. The data displayed here show a comparison of elemental peak areas to the area of Bremsstrahlung radiation in the range from 5.0 to 6.0 keV for variation in the elemental concentrations in frozen aqueous thin films from 1.5 mM to 25 mM. Peak areas in concentration of 1.5 mM are detectable above the background. The bar shows 90 % confidence limits.

Subcellular in-vivo concentration determination

The method of rapid freezing, i.e. a cooling rate of 10^4 deg C s^{-1} to the temperature of liquid nitrogen, effectively stops ion motion during cellular metabolic processes. If temperatures of less than -140 °C are maintained throughout the processing steps prior to observation, determinations can be made of the concentrations to a spatial localisation accuracy of roughly 10 nm provided adequate electron beam currents are available (approx.

5×10^{-10} A). These currents in a 10 nm probe are available only through the use of a field emission cathode; conventional thermionic sources are limited to diameters greater than 25 nm. Thus 'in-vivo' determinations of elemental concentrations can be made on a scale small with respect to the size of the structural elements present, as has been done in our laboratory in the case of skeletal muscle (Bacaner, Broadhurst, Hutchinson & Lilley, 1973). This capability truly opens up the area of fine structure elemental microanalysis determinations.

Present limitations

Elemental loss during analysis

Early studies by Bahr, Johnson & Zeitler (1965) showed that the lower the atomic number of the element, the more rapidly it is lost during electron, beam irradiation. Chemical bonding states will alter this general trend with strongly bonded elements being less susceptible to loss. Further work in our laboratory (Delgado & Hutchinson, 1975) and that of Hall at Cambridge (Hall & Gupta, 1974) has shown that these losses are significant during the time necessary for microanalysis to be performed. Recent studies (Hall & Gupta, 1974; Delgado & Hutchinson, 1975) have shown that these effects are minimised at very low temperatures. In addition, work is presently proceeding in our laboratory on quantification of the elemental loss during irradiation which should provide a means of calibrating the mass loss during analysis. The device used in this study consists of an ultra-high vacuum SEM fitted with a cryogenic specimen stage and a quadrupole mass spectrometer to quantify elements lost by known radiation doses. It may thus be possible to infer concentrations of elements at zero irradiation time from data obtained after radiation loss if the net electron dose and dose rate are known. An example of elemental loss curves for chlorine at 22 °C and −145 °C is shown in Fig. 5.

Specimen contamination due to 'hydrocarbon deposition' during analysis

Electron beam interaction with ambient hydrocarbons in the vacuum system of the electron microscope can lead to an increased net specimen thickness, due to the increase in hydrocarbon contamination. This is illustrated in Fig. 6, which shows in (*a*) the location of long-term 'microspot' analysis by STEM and in (*b*) a tilted SEM image of the hydrocarbon contamination peak which has been produced. The increased mass thickness decreases the apparent

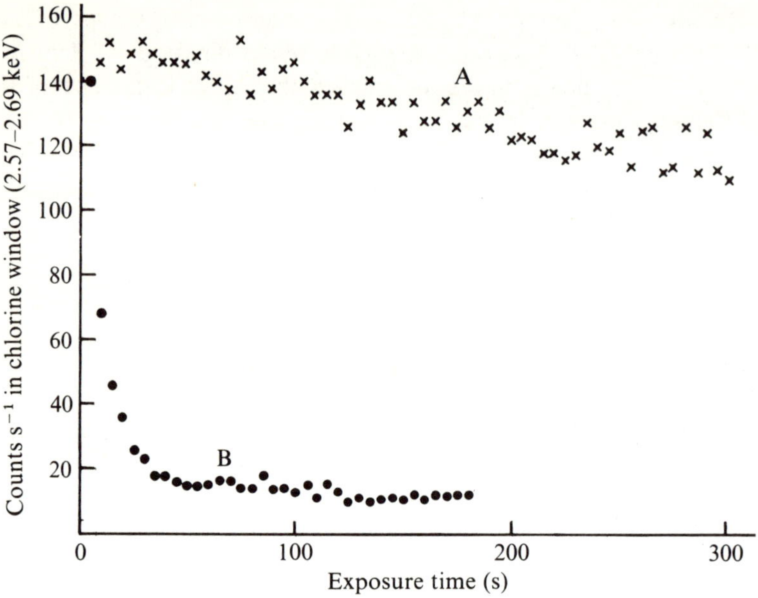

Fig. 5. The data shown here demonstrate the effect of temperature reduction from 22 °C to −145 °C on chlorine loss from a PVC thin film during electron beam irradiation at a beam current of roughly 5×10^{-11} A. A usable spectrum is obtained in 100 s at this current. A, PVC film mounted on 400 mesh copper grid at −145 °C with exposed area 2.58×10^{-9} cm² and irradiated with a beam current of 6.8×10^{-11} A. B, Unsupported PVC film at 22 °C with exposed area of 2.58×10^{-9} cm² irradiated with a beam current of 7.2×10^{-11} A.

concentration of all elements except those contained in the contamination, in this case carbon and trace sulphur, thus making quantification impossible unless the exact concentration of contaminant is known. Reduction of this problem to nil has been achieved with cold stages and well trapped vacuum techniques, but remains a problem for freeze-dried and embedded methods.

Spatial elemental migration during freezing

It has been shown by many investigators that if freezing rates are sufficiently slow, ice crystals are formed within the tissue, resulting in migration of ions during the process of freezing. If, however, the rate of freezing is sufficiently fast that amorphous ice is formed, no such migration occurs. Studies in our laboratories and others indicate that this rate of freezing must be of the order of 5×10^3 deg C s^{-1} or greater in order to preclude the possibility of ionic migration. Thus the heat conductivity of biological tissue limits the

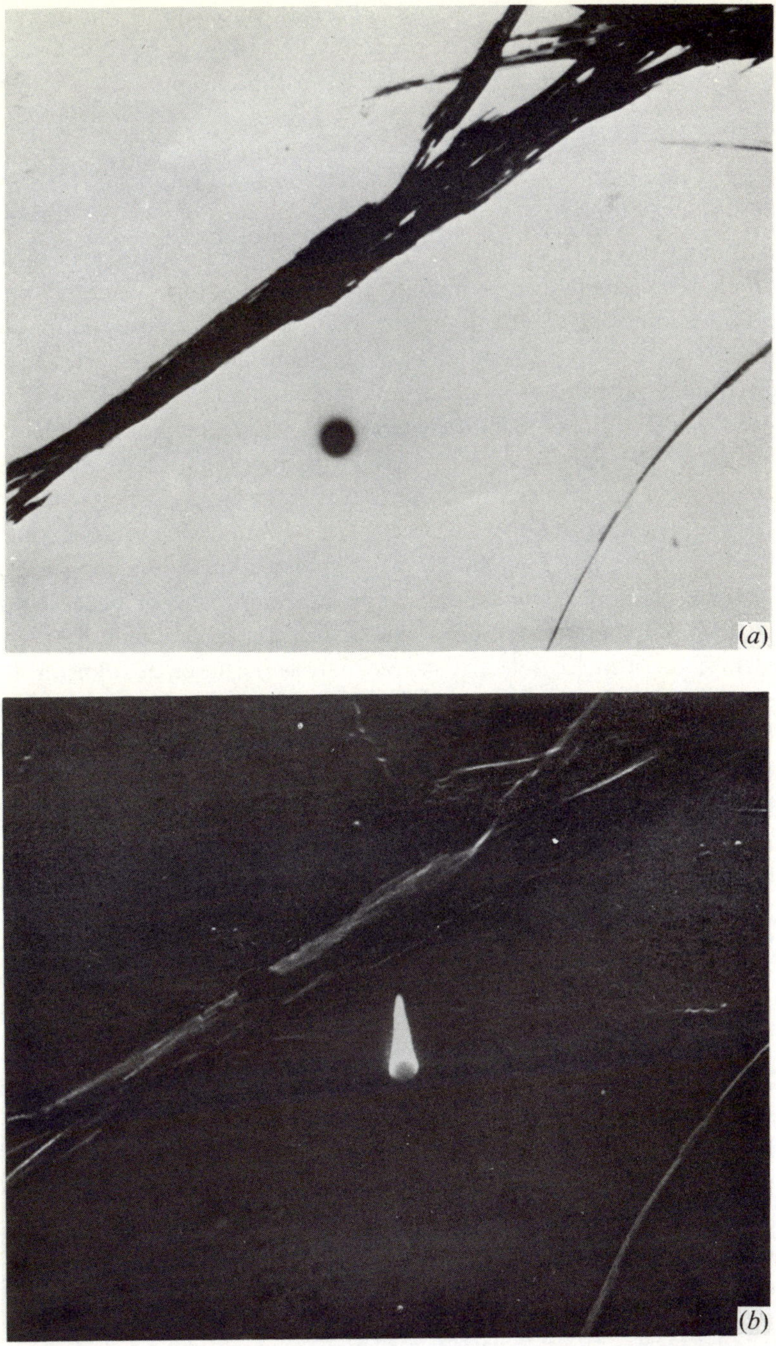

Fig. 6. This figure shows a transmission electron micrograph (*a*) and a secondary electron micrograph (*b*) of carbonaceous contamination formed during microspot irradiation of a thin carbon support film. The conical shape is clearly evident in (*b*).

thickness of the specimen to be successfully frozen to the order of some 30 μm. Studies are, however, proceeding in the use of cryo-protectants in order to preclude ice crystal formation in thicker specimens.

Elemental concentration detection limit and electron energy loss spectrometry

Although the energy selective X-ray analysis technique opens up vast new areas of microanalysis in biology the concentration detection limit is still a bothersome feature. Little future advance can be seen in refining the techniques to detect concentrations smaller than those of roughly 1 mM without an entirely new system of detection. This system of detection has presented itself in the form of electron energy loss spectrometry. This technique has as its core the detection and analysis, with very high efficiency, of energy loss electrons in the scattered electron beam from the specimen (for further discussion of this technique see Ferrier, this volume). Recent experiments with this technique have indicated that the detection limits for light elements includng carbon, nitrogen and oxygen are improved by at least a factor of 10 and frequently a factor of 1000 over those available to conventional techniques of energy selective X-ray analysis (Isaacson & Johnson, 1975). The employment of this technique should significantly reduce the limits of detection of the basic electron-excited elemental concentrations to approx. 0.01 mM, which again significantly advances the possibilities for determinations of ion movements within cells. Commercially produced systems are available although of unnecessary complexity and of decidedly inordinate expense. Systems are under development (D. Johnson, University of Seattle, personal communication) specifically tailored to the biologist which promise to be suitable for microanalysis of these low concentrations.

Microstructural localisation

The question of microstructural localisation previously mentioned is presently under investigation by dark-field techniques (Hutchinson, Johnson & Ris, in preparation). In addition, enhanced contrast in the STEM mode may well provide solutions to this problem (see Burge, this volume). Differential scattering by the various elements, and their natural concentrations, in unstained, unfixed material is the central feature of this difficulty. Operation of the instrument at low voltage of the imaging beam is a significant advantage in determining differences in concentrations of light elements.

Conclusion

Energy selective X-ray elemental microanalysis has in fact been demonstrated to be a most significant tool for the determination of concentrations of elements in volumes small with respect to that of a cell and is at present the only technique which allows this determination to be made. With the additional facility of the examination and analysis of rapidly frozen hydrated sections, this technique provides an in-vivo method for the determination of the concentration of individual ions and un-ionised elements. Although significant problems remain in the implementation of the technique, each problem has been outlined with a clear solution to the difficulties involved. These techniques do in fact present the possibility for a giant step to be taken in the area of microanalysis of elemental concentrations in biological tissue. Clearly the advance is possible through implementation of this technique in a careful and quantitative way and will allow the advance of our knowledge of the exact mechanisms of cell action through the concentration differences in elements during that action. No other technique at present provides the wealth of data afforded by the use of energy selective X-ray elemental analysis induced by electron beam irradiation.

I gratefully acknowledge the contributions of my graduate students to the data on microanalysis of biological material presented here, particularly Elizabeth Marston, David Stamerjohn, Louis Delgado and James Borek. I am also deeply indebted to the Kevex Corporation for many years of fruitful collaboration.

References

Bacaner, M., Broadhurst, J., Hutchinson, T. E. & Lilley, J. (1973). Scanning transmission electron microscope studies of deep-frozen unfixed muscle correlated wih spatial localization of intracellular elements by fluorescent x-ray analysis. *Proceedings of the National Academy of Sciences, USA*, **70**, 3423–7.

Bahr, G. F., Johnson, F. B. & Zeitler, E. (1965). The elementary composition of organic objects after electron irradiation. *Laboratory Investigation*, **14**, 1115–33.

Christensen, A. K. (1971). Frozen thin sections of fresh tissue for electron microscopy with a description of pancreas and liver. *Journal of Cell Biology*, **51**, 772–83.

Delgado, L. & Hutchinson, T. E. (1975). *National Science Foundation Report*.

Echlin, P. & Moreton, R. (1974). The preparation of biological materials for x-ray microanalysis. In *Microprobe Analysis as Applied to Cells and Tissues*, ed. T. Hall, P. Echlin & R. Kaufmann, pp. 159–175. Academic Press, New York & London.

Hall, T. A. & Gupta, B. L. (1974). Measurement of mass loss in biological specimens under an electron microbeam. In *Microprobe Analysis as Applied to Cells and Tissues*, ed. T. Hall, P. Echlin & R. Kaufmann, pp. 147–58. Academic Press, New York & London.
Hutchinson, T. E., Bacaner, M., Broadhurst, J. & Lilley, J. (1974). Instrumentation for direct microscopic elemental analysis of frozen biological tissue. *Review of Scientific Instruments*, **45**, 252–5.
Isaacson, M. & Johnson, D. (1975). The microanalysis of light elements using transmitted energy loss electrons. *Ultramicroscopy* **1**, 33–52.

J. A. CHANDLER

Wavelength dispersive X-ray microanalysis in biological research

X-ray microanalysis is a technique which allows a chemical analysis to be performed on biological tissue (and other material) within very small and well defined regions of the specimen. It makes use of the fact that atoms, when struck by electrons from an external source, yield X-rays which are characteristic of those atoms and are used to identify and quantify the elements present.

The idea of using electron-generated X-rays from atoms for identification was first proposed in 1943 but it was not until the late 1940s that it became a practical reality when the two techniques of X-ray spectrometry and electron optics were combined. Since then, and especially in the last 15 years, many hundreds of instruments employing X-ray microanalysis have been produced and used in such diverse scientific areas as metallurgy, physics, electronics, mineralogy, environmental pollution, geology, and lately in pathology, zoology, biochemistry and other biological fields. However, the number of instruments being used in biological research is a tiny proportion of the whole, due in part to the somewhat forbidding nature of the technology, and also to the peculiar physical and chemical properties of biological material.

The purpose of this paper is to enlighten the biologist who wishes to gain an understanding of, and greater confidence in, the principles of X-ray microanalysis and to help him determine the potential such a technique might have in his own field of research.

Several classes of problem can be tackled using the technique, as listed briefly below.

(i) *Natural elemental content of tissues.* The distribution of elements throughout subcellular regions can be demonstrated by analysis of normal physiological levels. As will be seen this requires a great deal of attention to be paid to methods of specimen preparation.

(ii) *Accidentally introduced foreign material.* The distribution of particles, toxic chemicals and strange elements can be located and identified within tissue.

(iii) *Deliberately introduced elements.* The results of administration of drugs and experimental substances can be traced throughout the tissue and related to morphological changes which occur therein.

(iv) *Histochemistry and immunochemistry.* This field is perhaps the most promising one for it allows the analysis of biochemical events *in situ.*

Although X-ray microanalysis, as defined above, allows the identification of individual elements, we are not always concerned with elements as such. Molecular groups and protein structures can be identified by specific labelling methods, e.g. by precipitating or substituting the molecular group *in situ* within the tissue, or by reacting protein in the tissue against specific antibodies to which enzymes and other labels have been coupled.

Other proteins and some smaller molecules such as steroids may also be labelled by coupling them in the tissue to heavy metals suitable as tracers for analysis by X-ray methods.

X-ray production

Characteristic radiation

Each element in the periodic table has a very well defined distribution of electrons within the atom. X-ray microanalysis is based upon the excitation of these electrons to produce an emitted X-ray spectrum which is characteristic of the element concerned.

In Fig. 1 the nucleus, comprising neutrons and protons, is surrounded by orbital electrons distributed quantum-mechanically in various energy levels. There may be many energy levels depending on the size and state of the atom and these are grouped into major units called shells. Elements are thus characterised by their nuclear charge, and number and energy distribution of orbital electrons. The element can be defined or identified by the electron energy levels or the energy difference between these levels. Large atoms (from heavy elements) contain greater numbers of electron orbits and shell units. If one of the orbital electrons is removed from its energy level by an incident electron then the atom is said to be in an excited state, or ionised. When this occurs an electron from a higher energy state will fall down into the gap to stabilise the atom. Because of the difference in potential energy levels the excess energy is emitted during this electron transition as an X-ray photon. This photon represents the exact difference in energy levels between the two electron shells. Thus if an electron in the K shell is removed, a second electron from the L shell may instantaneously replace it, thus giving off its excess energy as a photon of energy $E_L - E_K$, generally called K_α

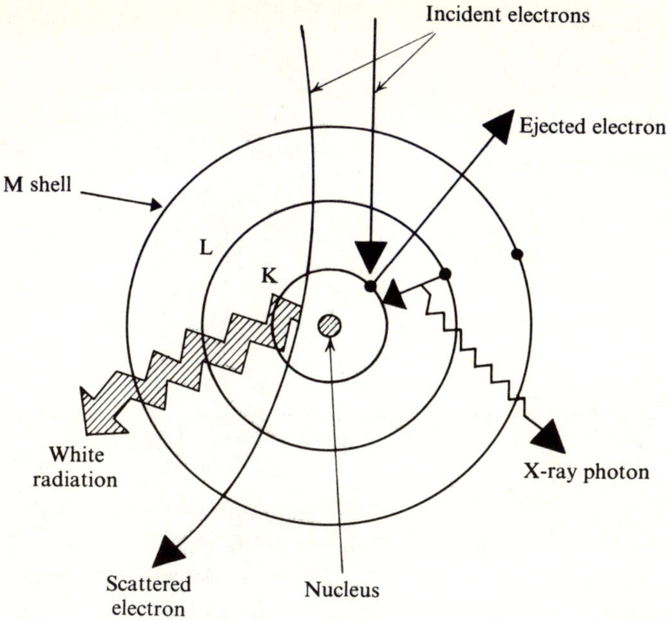

Fig. 1. Simple schematic representation of the atom, showing X-ray production by electron ionisation of an inner shell.

radiation. Now the filling of the vacancy in the K shell by an electron from the L shell will also produce a vacancy in the L shell. In turn this may be filled with one from the M shell and so on, each with the production of X-ray photons of energies determined by the orbital energies. Thus a single ionisation can give rise to a whole spectrum of characteristic X-rays and this energy spectrum identifies the atom.

X-ray continuum

Referring again to Fig. 1 a primary electron beam, instead of interfering with the orbital electrons to produce characteristic X-rays from the atom, may interact with the nucleus. As the incoming electron beam is decelerated by the field of nuclear charge it radiates energy. This loss of energy can be anything from the maximum originally carried by the electrons to a small fraction of it.

Fig. 2 shows the energy spectrum produced by this effect, with the maximum energy being that of the primary electron beam E_0. This general spectrum is called 'X-ray continuum' or 'continuous radiation', 'white radiation' or 'Bremsstrahlung'. It forms the background upon which the characteristic X-ray lines are superimposed at specific wavelengths (or

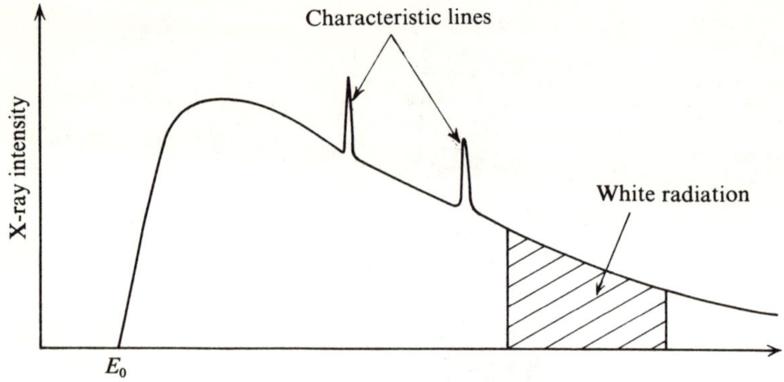

Fig. 2. Energy spectrum produced by electron interaction with matter. Characteristic lines are superimposed on the white radiation.

energies). This background must be subtracted from the characteristic signal produced by the specimen to obtain a valid reading of intensity. X-ray continuum forms a basic limitation to the ability to detect a characteristic 'line' since it can be considered as X-ray noise. In addition the continuum can cause fluorescence of X-rays from other atoms within the specimen. It can, however, be of great use in quantitation since it can provide a measure of specimen 'mass thickness' in order to determine elemental concentrations.

The X-ray signal collected from a given sample in a given time, under given experimental conditions, and for a particular element, may be considered to be proportional to the mass of that element. If the mass thickness of the specimen is also measured by collection of the white radiation (for example a region such as the hatched area of Fig. 2), then a method of calculating elemental concentrations is available. With crystal spectrometers, over a wide range of intensities, the X-ray signal is proportional to the number of atoms of a particular element in the area being analysed. With thick specimens precautions have to be taken regarding possible absorption as the X-rays leave the specimen, but this is not troublesome in thin specimens. Standards having known weight fractions of the elements concerned are used for calibration and such standards need to be close in nature to the specimen itself (Chandler, 1976).

Instrumentation

Several configurations of electron optical systems are available, many of which make use of X-ray analysis facilities. For biological materials the methods of viewing the specimen fall into two main categories. For thin specimens the image is viewed in transmission and for bulk specimens by reflection.

For biological work the instrumental arrangements of most interest are:

(*a*) scanning electron microscope + X-ray detection (SEM + X-ray);
(*b*) scanning transmission electron microscope + X-ray detection (STEM + X-ray);
(*c*) transmission electron microscope + X-ray detection (TEM + X-ray).

The X-ray detection systems most commonly used are: wavelength dispersive crystal spectrometers and energy dispersive solid-state detectors.

The two important parameters to be combined are the best resolution for imaging the specimen and the highest sensitivity available from the X-ray detector system.

Wavelength dispersive crystal spectrometer

Fig. 3 illustrates the principle of wavelength crystal spectrometers. X-rays leaving the specimen may cover a range of wavelengths from a range of elements excited by the electron beam. X-rays leave the specimen at all angles but a narrow cone passes towards a specially curved crystal such that a fraction of this signal is reflected into a detector. The fraction of the X-ray beam which is reflected depends upon the ability of the crystal to 'diffract' a particular wavelength while excluding all other wavelengths.

This phenomenon is described by Bragg's law which states that for crystals:

$$2d \sin \theta = n\lambda,$$

where n is an integer, λ is the X-ray wavelength successfully diffracted, d is the lattice spacing of the crystal and θ is the angle of incidence (and of reflection) of the X-ray beam at the crystal.

This law illustrates that, given a crystal of known lattice spacing d, and the angle of incidence θ, then the wavelength of the diffracted X-rays can be calculated. In practice the crystal is scanned through a range of angles until a 'peak' is noted in the detector at which point the conditions of Bragg's law are satisfied and the X-ray detected can be identified.

For good resolution and highly efficient diffraction of the X-ray lines, the crystal must be of excellent quality with no flaws or blemishes. The machinery that governs the position of the crystal relative to the specimen and to the detector must be of the highest accuracy to achieve reproducibility of results.

After being selectively diffracted by the crystal the X-rays of a particular

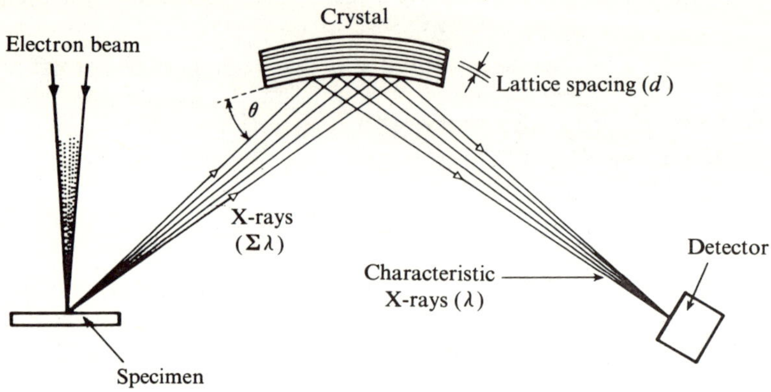

Fig. 3. Operation of a wavelength crystal spectrometer. The wavelength of X-rays successfully reflected (diffracted) depends on the angle of the crystal.

wavelength pass towards the spectrometer detector which is usually a gas flow, or sealed, proportional counter.

The range of wavelengths that such a crystal can cover depends upon the range of angles that it can be varied through. This limits a crystal to a few tenths of a nanometre of wavelength corresponding to the X-ray lines of few elements in the periodic table. To extend this range a diffracting spectrometer is often equipped with a number of crystals having different d values, or lattice spacings, so that for the same θ range the wavelength (λ) range is enlarged and thus a greater number of elements are covered.

Energy dispersive analyser

A limitation of the crystal spectrometer is that it is unable by definition to detect and display all X-ray energies leaving the specimen at once. The energy dispersive analyser, otherwise called a solid-state detector (SSD), provides an energy spectrum of all elements analysed simultaneously.

A very great advantage of the solid-state detector is that it can be placed very close to the source of X-rays and so accept a wide solid angle of radiation. This increases detection sensitivity enormously and thus provides better statistical data. Such detectors are discussed more fully elsewhere (Hutchinson, this volume), but they form such an essential part of any comprehensive analytical system that they will be included in this paper.

The type of system chosen for a particular application depends entirely on the nature of the problem. When dealing with biological materials, however, and except in the case of teeth or bone, the elemental levels are frequently very low ($< 1\%$) and special conditions govern the choice of analysing system as well as electron optical system.

Energy resolution

Energy resolving power of a detector may be defined as the ability to separate two adjacent peaks in the energy spectrum. The resolution is generally taken to be the full width of the peak at half the maximum intensity (fwhm). Besides major X-ray lines, there are many other overlapping lines in the energy spectrum when a number of elements are present together in the specimen region being analysed. For example the K_α line of phosphorus occurs at 2.01 keV whereas the M_α line of osmium, a frequently employed stain in electron microscopy, occurs at 1.91 keV, only 100 eV away. Such overlaps occur frequently in biological studies, especially when heavy metals are employed to provide stains or are used in cytochemical techniques. The ability to resolve two such lines depends on the relative peak heights, i.e. relative concentrations of the elements. Heavy metal stains are frequently present in concentrations much greater than the naturally occurring elements and resolution requirements vary.

Solid-state detectors currently commercially available have resolving powers of about 160 eV, although this figure is being improved continually. The resolution can also be artificially increased (made better) by computer techniques of spectrum stripping where the noise and statistical fluctuation are smoothed out. Physical limitations however (thermal and quantum noise) place a lower limit on the resolving power likely to be achieved with this detector (possibly around 100 eV).

Crystal spectrometers function at a much higher resolution by the nature of their operation. The technique of wavelength dispersion by diffraction ensures that only those X-rays of a very precise wavelength (or energy) reach the detector. Resolving powers better than 10 eV can be achieved routinely with such a system.

Sensitivity

The ability to resolve X-ray lines alone is insufficient unless the detector system allows adequate X-ray counts to be accumulated in reasonable analysis times. However, a distinction must be made here between 'count rate sensitivity' and 'mass detection sensitivity'.

The solid-state detector has the great advantage of being able to be placed very close to the specimen (sometimes just a few millimetres away). This vastly raises the count rate. Thus a greater number of X-ray photons can be detected in the same time or the same number in a shorter time. This means that a lower electron signal can be employed to generate a greater X-ray signal, thus allowing a smaller probe diameter and reducing specimen damage. However, a faster counting rate does not necessarily in itself increase mass detection sensitivity since this depends critically on the ratio of peak counts to background counts, for which the wavelength dispersive crystal spectrometer is much superior.

To make a direct comparison between a solid-state detector and a crystal spectrometer arrangement would not be helpful because of the very wide variation in design. Either system may be used with the electron optical arrangements described below and often they are used in conjunction. The wavelength dispersive crystal spectrometer and the solid-state detector essentially provide the same information about the elemental nature of the specimen but present it in different ways.

The various combinations of electron optical systems with X-ray microanalytical detectors that are of most interest to cytochemistry are illustrated below.

Scanning electron microscopy + X-ray analysis

Fig. 4 illustrates the arrangement by which crystal spectrometers or solid-state detectors, or both, may be incorporated into a scanning electron microscope. The microscope may be operated for secondary emission or transmission imaging. A great advantage of the SEM is the ease with which detectors may be brought close in to the specimen for greater sensitivity.

The SEM allows X-ray information to be obtained in the form of a two-dimensional elemental map across the specimen, as a line trace across a region of interest, or by static probe analysis.

Two dimensional analysis is performed by setting the detector to collect only the particular element of interest and then allowing the focused electron beam to scan the specimen surface. The X-ray signal from the detector is displayed on a cathode ray oscilloscope by synchronising with the electron beam raster.

For one-dimensional analysis, i.e. a line trace, the output is simply fed into a chart recorder so that changes in elemental content can be monitored across that line in the specimen.

Most sensitive and quantitative analysis is obtained with a static probe positioned over the region of interest and by counting the X-ray emission over a period of time. For qualitative analysis the crystal spectrometer is made to run through a range of wavelengths to cover the elements of interest.

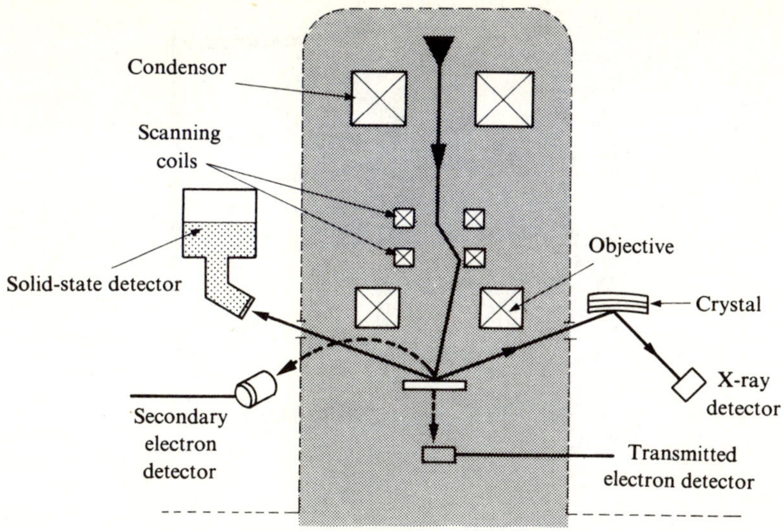

Fig. 4. Schematic representation of a scanning electron microscope incorporating X-ray microanalysis.

With the solid-state detector, qualitative information is at once produced as an energy spectrum for the static probe.

Transmission electron microscopy + X-ray analysis

A logical extension of transmission electron microscopy is to combine high-resolution imaging of ultrathin sections with high-efficiency X-ray microanalysis.

Fig. 5 illustrates this combination in the integral instrument EMMA (electron microscope microanalyser) where the design has been such as to avoid compromise between the high-resolution electron imaging and the efficiency of analysis (Chandler, 1975). The departure from a conventional electron microscope exists in the centre where, immediately above the specimen an extra condenser lens is involved. This small conical lens, called a 'minilens', allows the illumination to be focused to a small probe on the specimen. Probe diameters of 0.1–0.2 μm are commonly employed.

Because of the nature of the imaging, the transmission image is visible on the microscope screen even during analysis so that very accurate location is possible. X-rays leaving the specimen enter the detector systems (double crystal spectrometers and solid-state detector) via thin windows to allow the microscope to operate at high vacuum and without introducing contamination into the specimen region. Both detector systems can function simul-

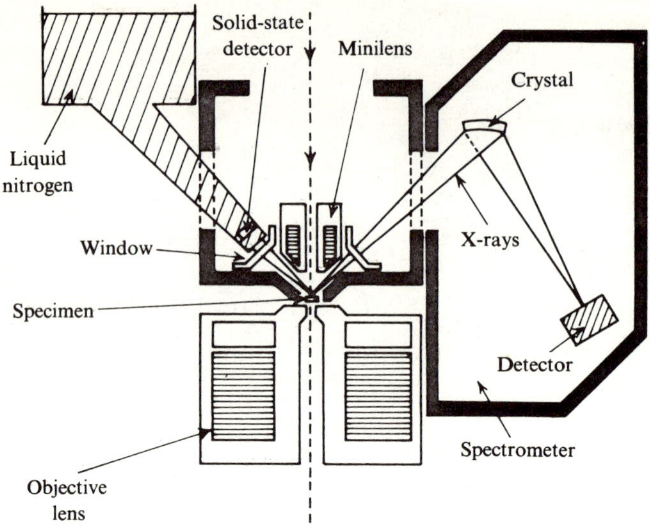

Fig. 5. EMMA – the electron microscope microanalyser, combining wavelength crystal spectrometers with a solid-state detector.

taneously, and in some arrangements the solid-state detector can be made to slide in and out towards the specimen to increase the solid angle of collection of X-rays.

Fig. 6 illustrates the arrangement and demonstrates the great advantage of this configuration whereby the ultrastructure can be correlated with subcellular microanalysis for detailed cytochemical studies.

The instrument is operated basically in the same mode for microanalysis as for normal transmission electron microscopy with the range of voltages from 40–100 kV suitable for most types of ultrathin specimen. Although beam deflectors are incorporated in the instrument in the condenser system, the probe is not normally scanned as in the previous configuration but the instrument is used essentially as a static probe analyser. With typical ultrathin sections ($<$ 100 nm) the amount of material to be analysed is extremely small (generally $\sim 10^{-15}$ g) and such scanning systems would rarely yield useful information. Typical counting times are in the region of 100 s per point analysis.

There are many hundreds of transmission electron microscopes in all corners of the world which are capable of being converted to X-ray analysis instruments simply by attaching a suitable detector to the specimen region. This often requires some modification to the microscope column but quite a satisfactory arrangement can often be obtained without too much disturbance of the electron optics.

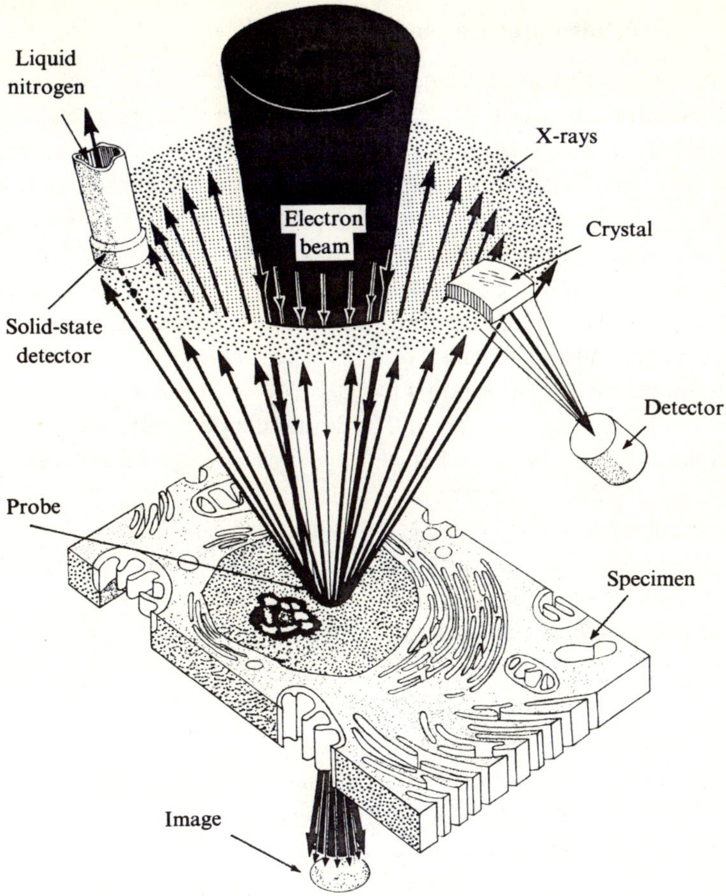

Fig. 6. Schematic representation of X-ray microanalysis of thin sections with transmission electron microscopy.

Crystal spectrometers were the first type of detector to be incorporated into microscopes in this way but these have to some extent given way to solid-state detectors with their rapid development in recent years. Such detectors are more versatile and flexible so that they can be interfaced between different electron optical instruments. There should be a great increase in the numbers of transmission electron microscopes having this attachment facility.

Again, like EMMA, this combination instrument finds its greatest use in static probe analysis and combines X-ray analysis with detailed morphological studies on ultrathin sections.

Specimen preparation

X-ray microanalysis has been performed on both hard and soft tissue in thick and thin forms as well as in isolated cells. For an interpretation of the analytical results the investigator must have full confidence in the integrity of the specimen, i.e. that elements that have been analysed shall be distributed within the specimen in the same way as in the original living tissue, or at least that any differences should be entirely understood. Different preparatory techniques will affect elements in the tissue in different ways depending on whether they are free or bound. Soluble electrolytes, for example, are unlikely to remain in their in-vivo sites when subjecting tissues to organic or aqueous solutions. The binding forces involved in linking elements to organic structures are very complex and little understood. Elements may exist also in tissue in both the bound form and as free ions and will be affected differently by different procedures.

Correlative techniques are necessary to establish the nature of these binding forces and the effects of various treatments upon them. For example, atomic absorption spectrometry, or radioactive tracer elements may be used to monitor the retention of elements within tissue samples during processing.

Thick specimens ($> 0.2\ \mu m$)

Liquid methods. Some standard histological methods may be quite suitable for some types of analyses provided one is confident that the final specimen is sufficiently similar to the original tissue. Such methods may involve the use of fixatives such as alcohol, aldehydes, formalin, or acetone; embedding media such as paraffin, styrene, methacrylate and low viscosity resins; clearing solutions such as xylol, toluol, benzene, chloroform or cedarwood oil; and possibly flotation on water and staining for histological correlation or histochemical tests.

All or any of these procedures can possibly contaminate the specimen or cause elemental loss or redistribution. Several cases have been quoted where this has arisen and the losses have been intolerable, yet other examples, on different tissues and with analysis of different elements have proven otherwise.

Freeze-drying. The commonly used method of rapidly freezing the tissue block, freeze-drying, vacuum impregnation and sectioning critically reduces the risk of sample contamination, loss of soluble substances, displacement of cell constituents and chemical alteration of reactive groups.

However the possibility of changes occurs during the freeze-drying procedure as the ice-vapour front moves through the tissue, and of contamination from the embedding medium where elements such as sulphur, chlorine, calcium and silicon may be present.

Cryomicrotomy. The technique of cutting frozen sections and freeze-drying them ready for analysis is of value in order to be absolutely sure that there is no risk attached to elemental redistribution or loss. Histological detail in these sections suffers from difficulties associated with ice-crystal formation, the hygroscopic nature of the specimen, unevenness, compression artifacts, and folds. The morphological information is nowhere near as good as with conventional methods but in many cases is good enough for the spatial resolution of analysis. With section thicknesses between 1 and 5 μm electron diffusion is considered to occur within a narrow cone, thus providing reasonable spatial resolution with relatively large quantities of material for analysis (compared with ultrathin sections).

Air-drying. For the analysis of cell suspensions (e.g. blood, sperm) relatively simple air-drying procedures can be adopted. Some care has to be taken to obtain 'pure' cells by this method. For example, human semen contains much debris in the seminal fluid which contaminates the cells. Washing the cells is not satisfactory because of elemental redistribution. Freeze-drying of cell suspensions is an alternative to air-drying.

Mounting. Specimens are mounted on flat substrates of materials that will not interfere with the analysis (e.g. carbon, quartz, beryllium, nylon film) and are thoroughly coated with a thin film (e.g. carbon or aluminium) to prevent electrostatic charging or specimen damage from the electron beam.

Thin specimens ($< 0.2\ \mu m$)

The same principles apply regarding elemental loss and redistribution as described for thick specimens but the methods involved in overcoming these difficulties vary somewhat. Since the principle of performing analysis on ultrathin sections is that of correlating ultrastructure with analysis, methods must be chosen which will not affect the chemicals of the sample too much but will also preserve the ultrastructure to a very high degree. Sensitivity of analysis is all important because of the relatively small amount of material compared with thick sections, but image resolution is far superior.

Liquid methods. An important property of the fixative is that it should form stable bonds which will hold the tissue molecules together during dehydration and subsequent treatments. Contamination may arise from buffers containing phosphorus, sodium or other elements. Dehydrating agents like alcohols, acetone and gelatin, as well as being agents for replacing water, are nearly all strong organic solvents which can cause some extraction of cell constituents. Propylene oxide has been found to extract lipids, possibly even from fixed tissue. The various embedding media may contain contaminating elements, and might introduce errors into the analysis of concentrations depending upon their stability in the electron beam. Sections need to be thin enough to allow good resolution in the image yet thick enough to allow sufficient quantity of elements for detection. Flotation on a water bath is another possible source of artifact where subcellular constituents may be displaced from the sections.

Staining. Electron stains increase the possibility of obtaining cytochemical information from the sections but can interfere with the analysis. The heavy metals employed produce characteristic X-ray emissions which may mask lines of other elements being examined. In some cases it is a quantitative estimation of the stain material itself which is required. Heavy metals also make the detection of trace elements within the tissue more difficult by raising the background signal.

Cryo-ultramicrotomy. The use of wet chemicals may be avoided altogether by cutting frozen sections for analysis. Analysis has been performed mostly on sections at room temperatures but efforts are being made to maintain them frozen and with the water still present during analysis in the microscope. The technique is very young at the present time and a great deal of work is yet to be done to improve the quality of ultrastructure obtained by this method. Cryo-ultramicrotomy is the topic of the chapter by Appleton (this volume).

Other methods of histochemistry using frozen fixed and freeze substituted tissue go some way to overcoming the losses involved in conventional methods.

Mounting. As with thick specimens, sections are mounted on supports that will not interfere with the analysis. Grids of carbon, beryllium and aluminium are commonly used. Sections are carbon-coated to overcome the damaging effects of the electron beam.

Applications

Whereas most applications of X-ray microanalysis as a whole have been on mineralised tissues such as teeth and bone, our major interest here is in subcellular analysis and we shall turn our attention to those applications in particular. Once again it is helpful to consider applications in terms of thick and thin specimens.

Thick specimens ($> 0.2\,\mu m$)

The distribution of calcium, iodine and phosphorus has been studied in human thyroid glands by Robison, Middlesworth & Davis (1971) using a scanning instrument with a crystal spectrometer. Specimens were prepared for analysis by freezing the tissue in liquid nitrogen and cutting 12-μm sections on a cryostat. Other samples were fixed in formalin and cut frozen or embedded and sectioned. Scanning X-ray images indicated a heavy distribution of calcium in follicular cells and colloid material, while iodine was found uniformly distributed in the follicles of normal thyroids.

Some of the more commonly used histological staining methods can be used in conjunction with X-ray microanalysis where detectable elements are found to be associated with the stain. For example, Gomori's chrome-haematoxylin could be identified by analysis of chromium in the tissue. Gallocyanin-chromalum is a chromium-containing dye that reacts with nucleic acids and has been used for quantitative determinations.

Since the need for an electron-dense final reaction product is obviated by the use of X-ray microanalysis this has important consequences in enzyme histochemistry. Engel, Resnick & Martin (1968) have examined 10-μm thick sections of fresh frozen human skeletal muscle to demonstrate alkaline phosphatase wihout going to the final reaction product of cobalt sulphide. Instead the distribution of calcium was observed by an X-ray scanning image after calcium hydrogen phosphate precipitation by the Gomori technique.

Thin specimens ($< 0.2\,\mu m$)

Trace elements in the reproductive system and their dependence on endocrine status have been studied using an EMMA instrument with ultrathin ($< 0.1\,\mu$m) sections. Specimens are fixed with potassium pyroantimonate and osmium tetroxide to retain cations *in situ*, which are then analysed quantitatively in each subcellular region. Fig. 7 illustrates a section through unstained epithelial cells of rat prostate acini cultured *in vitro* and fixed in this manner. The distribution of antimonate deposits is evident. The effect of various hormonal controls on zinc, cadmium and calcium distributions can be successfully measured in nucleoli, nucleoplasm, secretory granules and other subcellular regions. The nucleus is found to contain high levels of zinc as are the secretory granules, presumably ready for secretion into prostatic fluid. A high concentration of stored zinc is found in the stromal elastin (Chandler, Timms & Morton, 1977).

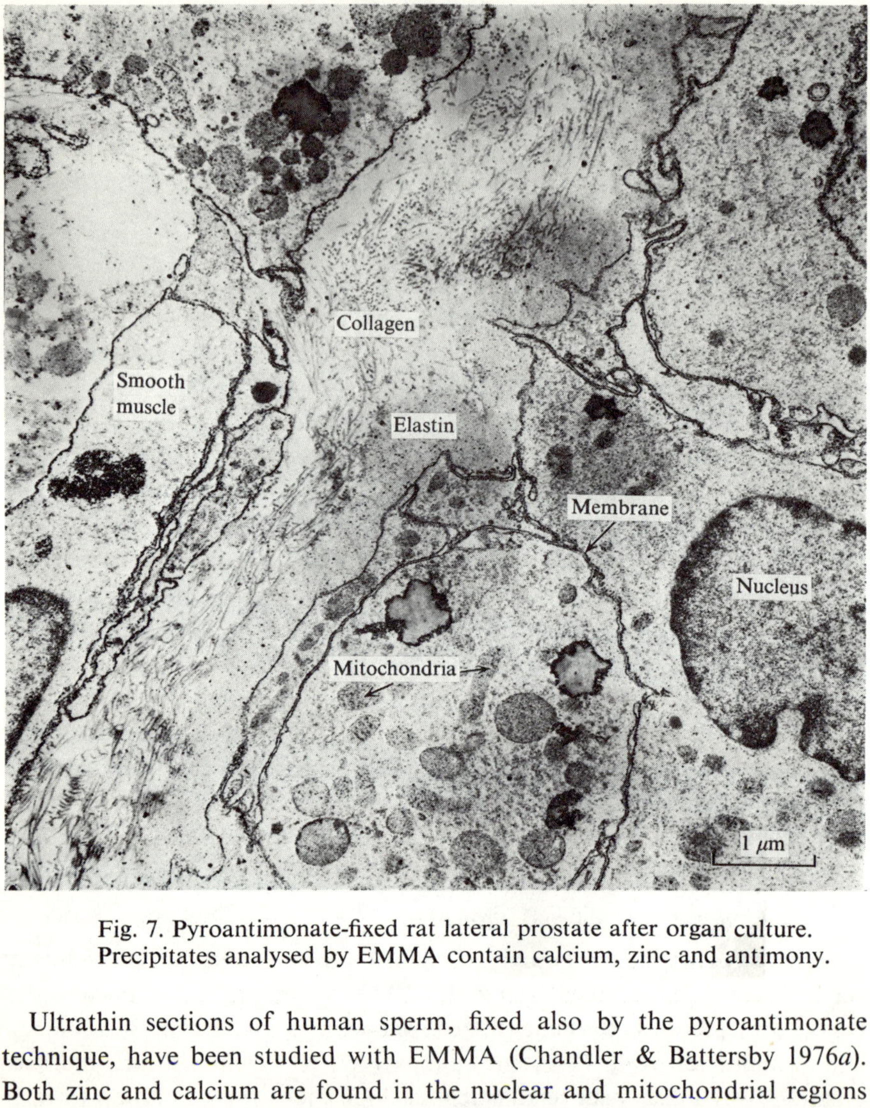

Fig. 7. Pyroantimonate-fixed rat lateral prostate after organ culture. Precipitates analysed by EMMA contain calcium, zinc and antimony.

Ultrathin sections of human sperm, fixed also by the pyroantimonate technique, have been studied with EMMA (Chandler & Battersby 1976*a*). Both zinc and calcium are found in the nuclear and mitochondrial regions (Fig. 8), often not being associated with antimony deposits and suggesting a firmer binding of these elements to the cell matrix.

Ultrathin frozen and freeze-dried sections of human sperm have been analysed and compared with air-dried whole cells. No differences were noted for the two methods of preparation for sodium, phosphorus, sulphur, chlorine, calcium, potassium or zinc in the subcellular regions (Chandler & Battersby, 1976*b*).

Electron histochemistry is still limited compared to the methods available

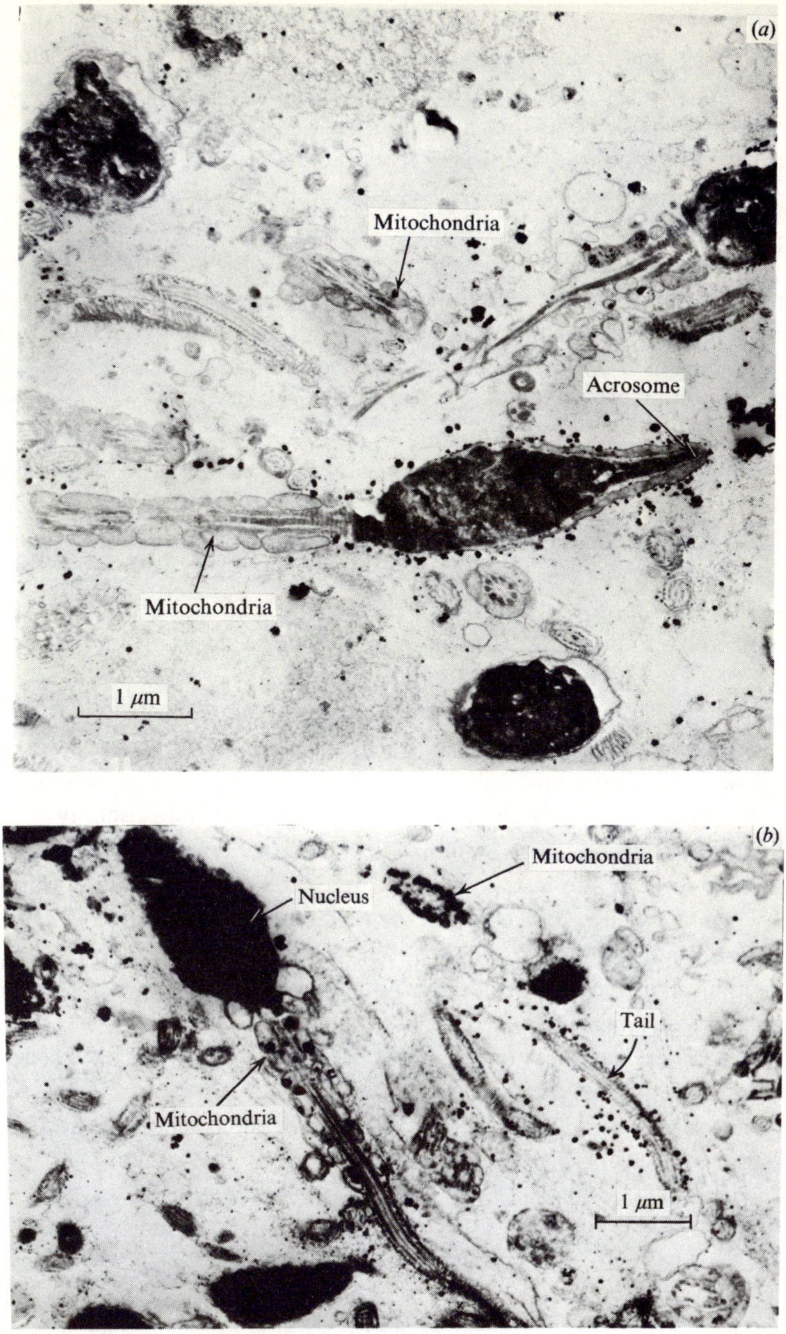

Fig. 8(*a*) and (*b*). Ultrathin section of human sperm cells fixed in pyro-antimonate. Zinc is found by analysis in the unstained nucleus and in precipitates in the mitochondria together with calcium.

for light microscopy since it depends on the production of an electron-scattering final reaction product. However X-ray microanalysis can be used effectively to extend the range of histochemical techniques by the localisation of individual elements that may constitute part of a stain that is not electron-dense. For example the ferric ferricyanide method for cystine-containing proteins could be employed where the presence of iron would define disulphide groups only.

Chromium compounds have been studied as stains for nucleic acids in ultrathin sections for electron microscopy. X-ray microanalysis could well be applied together with those techniques for quantitative nucleic acid determinations *in situ.* Other methods for staining nucleic acids may also be used in microanalysis, involving the metals silver, lead, uranium, indium, bismuth or mercury.

An extremely important application is the labelling of steroids with elements detectable by microanalysis. Some steroids, notably oestradiol, have an affinity for mercury. For example the 17β-oestradiol can be reacted with a mercurial salt to form the 4-mercuri-17β-oestradiol. Aromatic mercurials have been found to react with sulphydryl groups which in turn have been implicated in the binding of 17β-oestradiol by receptor protein. Thus mercury may be used in this way for affinity labelling studies of receptor protein in subcellular regions of thin sections. It is thought that strong mercaptide binding will occur at the receptors such that wet chemical preparatory methods may be employed.

For immunocytochemistry X-ray microanalysis may usefully be used for quantitative studies of protein in thin sections. Ferritin is easily measured by its iron content and in the method employing acid phosphatase the lead ions are readily analysed. However, the advantages of the peroxidase technique are so important for demonstrating antibody in tissue that methods of forming final reaction products containing detectable elements would be of great value (e.g. labelling diaminobenzidene (DAB)).

Many other applications of X-ray analysis to thick and thin specimens of biological material are given by Hall (1971) and Chandler (1975). A detailed review of all techniques used for X-ray microanalysis in the electron microscope is given by Chandler (1977).

Conclusion

X-ray microanalysis, combined with electron optical imaging of thick or thin specimens, thus has enormous value in the examination of biological material. Recent developments in instrumentation have provided the possibility of detecting elemental levels down to 10^{-18} or 10^{-19} g and local weight fractions of 50–100 ppm within subcellular organelles. Success in the use of the technique depends on a good understanding of the behaviour of tissue elements during specimen preparation and on a careful control of instrumental and operating conditions during the analytical procedure. Wavelength dispersive crystal spectrometers have largely given way to energy dispersive solid-state detectors in recent years, largely because of the convenience of use of the latter and the relative ease with which they can be retrospectively fitted to existing microscopes. When trace elements are to be detected in biological materials, however, and when these specimens may present complex elemental spectra during analysis, the advantages of the crystal spectrometer become very important.

The author is grateful to the Tenovus Organisation for generous financial support.

References

Chandler, J. A. (1975). Electron probe X-ray microanalysis in cytochemistry. In *Techniques of Biochemical and Biophysical Morphology*, ed. D. Glick & R. Rosenbaum, vol. 2, pp. 308–437. Wiley, New York.

Chandler, J. A. (1976). A method for preparing absolute standards for quantitative calibration and measurement of section thickness with X-ray microanalysis of biological ultrathin specimens in EMMA. *Journal of Microscopy*, **106**, 291–302.

Chandler, J. A. (1977). X-ray microanalysis in the electron microscope. In *Practical Methods in Electron Microscopy*, ed. A. M. Glauert. North-Holland Publishing Co., Amsterdam.

Chandler, J. A. & Battersby, S. (1976*a*). X-ray microanalysis of zinc and calcium in ultrathin sections of human sperm cells using the pyroantimonate technique. *Journal of Histochemistry and Cytochemistry*, **24**, 740–8.

Chandler, J. A. & Battersby, S. (1976*b*). X-ray microanalysis of ultrathin frozen sections of human sperm cells. *Journal of Microscopy*, **107**, 55–65.

Chander, J. A., Timms, B. G. & Morton, M. S. (1977). Subcellular distribution of zinc in rat prostate studied by X-ray microanalysis. I. Normal prostate *Histochemical Journal*, **9**, 103–20.

Engel, W. K., Resnick, J. S. & Martin, E. (1968). The electron probe in enzyme histochemistry. *Journal of Histochemistry and Cytochemistry*, **16**, 273–5.

Hall, T. A. (1971). The microprobe assay of chemical elements. In *Physical Techniques in Biomedical research*, 2nd edn, vol. 1A, pp. 157–275. Academic Press, New York & London.

Robison, W. C., Middlesworth, L. V. & Davis, D. (1971). Calcium, iodine and phosphorus distributions in human thyroid glands by electron probe microanalysis. *Journal of Clinical Endocrinology and Metabolism*, **32**, 786–95.

T. C. APPLETON

The use of ultrathin frozen sections for X-ray microanalysis of diffusible elements

Biologists have traditionally used the microscope as a means of studying the organisation and structure of cells within tissues, organs or cell suspensions. While the electron microscope has enabled such studies to be extended to the level of cell organelles (and under ideal conditions to molecules) the emphasis in the past has mainly been on structure rather than function. The advent of sensitive X-ray analytical techniques has meant that the biologist is now able to turn his attention to the chemical nature of the cells and their constituents. The optical and analytical levels of resolution which can now be attained readily allow areas as small as 25 nm diameter to be analysed if the elemental concentration is sufficiently high (e.g. in 1–2-μm sections of mineralised inclusions), or in the region of 100 nm diameter where concentrations are low (e.g. in 100-nm sections). It is probable that X-ray microanalysis of still smaller areas will soon be possible with such a rapidly advancing technology.

The majority of elements of biological interest are present in a highly diffusible form – e.g. sodium, magnesium, phosphorus, sulphur, chlorine, potassium, calcium, iron – although some may be bound as insoluble or partially insoluble forms. Elements which are *known* to be bound into biological materials may be studied after conventional fixation and embedding provided that (*a*) there is sufficient evidence to suggest that material is not lost during the preparative procedures – loss of material, however small, suggests translocation of the remaining elements; and (*b*) fixatives and stains which contain heavy elements are only to be used if the energies of their X-rays do not obscure the elements of biological interest.

The majority of elements of physiological interest are highly soluble or only lightly bound and are lost or redistributed during conventional fixation, dehydration or embedding procedures. It is therefore necessary to use preparative procedures which retain all the diffusible components of the cell; loss or diffusion of one element suggests that other elements may have been redistributed. Several techniques are now available for cutting ultrathin frozen sections of biological material. Sections can be cut from fixed or unfixed

Table 1. *Techniques for making 'dry' ultrathin frozen sections*

(1)	Freeze 1-mm^3 pieces or droplets in liquid nitrogen slush
(2)	Store in liquid nitrogen
(3)	Mount in cryostat and trim to pyramid (−70 to −90 °C)
(4)	Draw ribbons using vacuum (50–100 nm)
(5)	Mount on to Formvar-coated grids using light pressure from copper rod
(6)	Freeze-dry in cryostat; 3 h at −80 °C
(7)	−80 °C to ambient temperature in dry nitrogen atmosphere or transfer directly to cold stage
(8)	Carbon-coat (100 nm+) under vacuum

material on to suitable flotation fluids such as dimethyl sulphoxide (Bernhard, 1971; Sjöström, Thornell & Cedergreen, 1973) or cyclohexene (Hodson & Marshall, 1970), or they can be cut on to dry glass or diamond knives (Appleton, 1968, 1969, 1972*a*, *b*, 1974*a*, *b*; Christensen, 1969, 1971; Sjöström & Thornell, 1975).

The purpose of this paper is (*a*) to describe the technique of cutting ultrathin frozen sections of unfixed, unembedded sections using a cryostat equipped with an ultramicrotome and capable of maintaining a temperature of −50 °C to −140 °C, (*b*) to evaluate critically the suitability of such sections when used in conjunction with X-ray microanalysis, and (*c*) to suggest some applications of these techniques.

The techniques are summarised in Table 1, and described in detail elsewhere (Appleton, 1974*a*, *b*).

Techniques

Freezing

The requirements of the freezing process are (*a*) to 'lock' the materials within the cells so that compartments of diffusible elements are retained, and (*b*) to retain sufficiently good morphological preservation of the cell and its constituents so that the elements can be localised to indentifiable structures.

There is no freezing method available which provides perfect ultrastructural morphology (Christensen, 1971); some ice-crystal damage is always present. The use of cryo-protective agents (e.g. DMSO or glycerol) must for the present be avoided since they are likely to induce the movement of water and solutes across membrane boundaries. Rapid freezing of small samples gives preservation with minimal ice-crystal damage; large areas can be examined in the electron microscope with sufficient contrast and detail to identify cells and their component parts.

Small pieces of fresh tissue (1 mm^3 or less), or small droplets of cell suspensions, are frozen directly on to brass or silver specimen holders by immersing the tissue in liquid nitrogen 'slush'. Liquid nitrogen (−196 °C) will freeze when placed in a rough vacuum (0.05 torr) for 2 min; when the vacuum is broken (i.e. air admitted) the solid nitrogen partially melts giving liquid nitrogen slush (−210 °C). Tissues immersed in the slush freeze very rapidly because the temperature is well below the boiling point of nitrogen. The slush is stable for 6 min under normal laboratory conditions.

Tissues frozen in this way can be stored in a liquid nitrogen refrigerator until required; there is no detectable deterioration of morphology or redistribution of diffusible substances over periods up to 3 years.

Sectioning

Sections are cut in a cryostat equipped with an ultramicrotome (Fig. 1). The cryostat described previously (Appleton, 1974*a*, *b*) has been modified so that the cooling of the chamber is by injection of liquid nitrogen through jets positioned at the base of the cryostat chamber: a compressor is only used to provide a stand-by temperature of −40 °C. Liquid nitrogen cooling provides a controllable dry environment free from vibration and with a temperature range from −50 to −140 °C. The temperature is controlled by varying the amount of liquid nitrogen injected into the chamber. Tests have shown that the air, knife and specimen temperatures are identical (±1 deg C) and that the chamber air is free from turbulence. A cryostat which incorporates these features is available from SLEE Medical Equipment Ltd.*

The frozen tissue is trimmed in the cryostat to a pyramid with a square or rectangular block face; the temperature during trimming should be similar to that used for sectioning (−75 to −90 °C depending on the tissue). Serial sections, cut with the advance mechanism set at 100 nm, are pulled away from the knife edge using a vacuum device (Fig. 1) mounted on a micromanipulator: the method of pulling a ribbon of ultrathin sections from a dry glass knife is shown in Fig. 2.

Ribbons of ultrathin frozen sections are pressed into contact with Formvar-coated support grids using light pressure from a cold flat polished copper rod (Christensen, 1969, 1971).

* Slee Medical Equipment Ltd, Lanier Works, Hither Green Lane, London SE13 6QD.

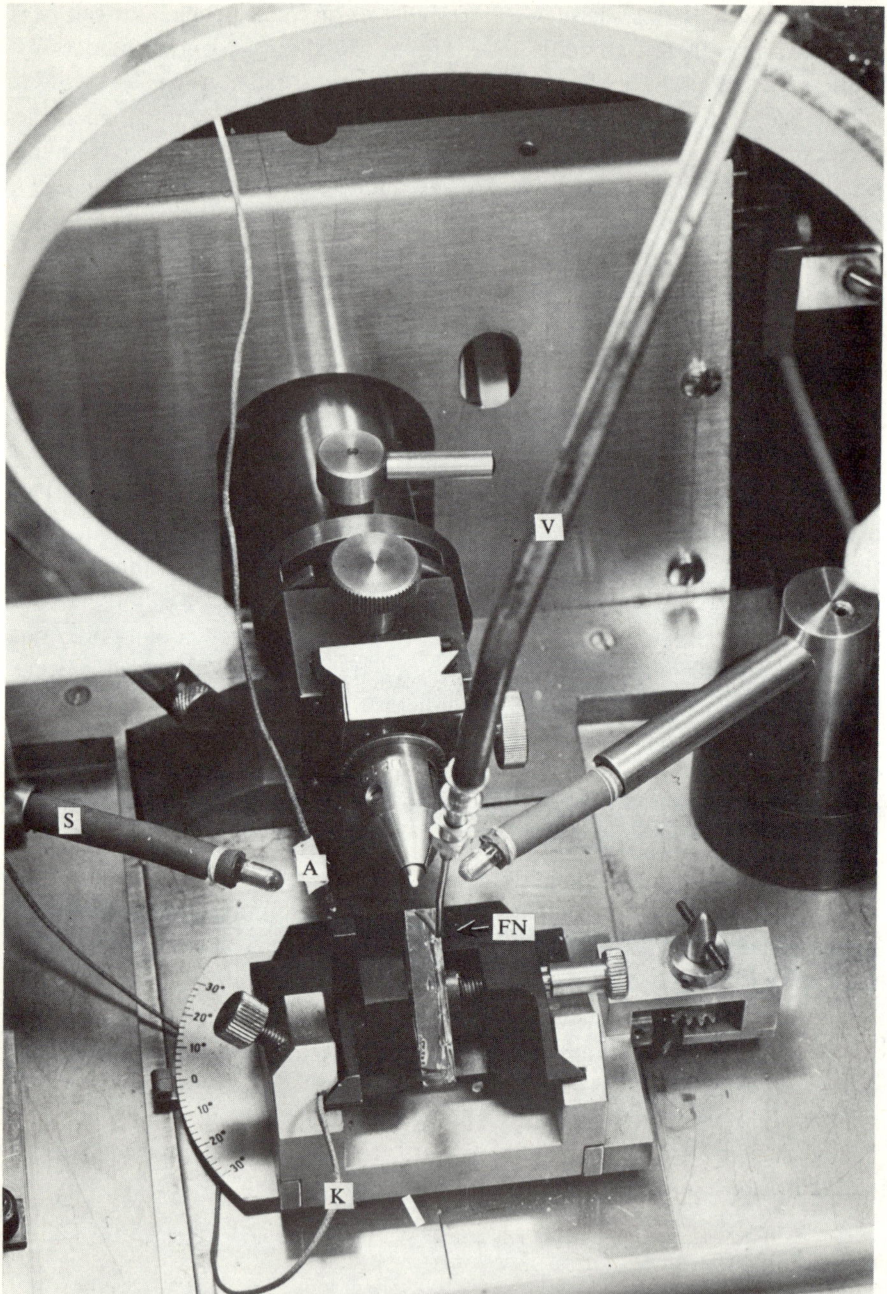

Fig. 1. Internal view of cryostat showing: V, vacuum device with flattened needle (FN) behind the cutting edge; S, sensors for thermostat; A, thermocouple reading air temperature; K, thermocouple for knife temperature. (After Appleton, 1974*a*.)

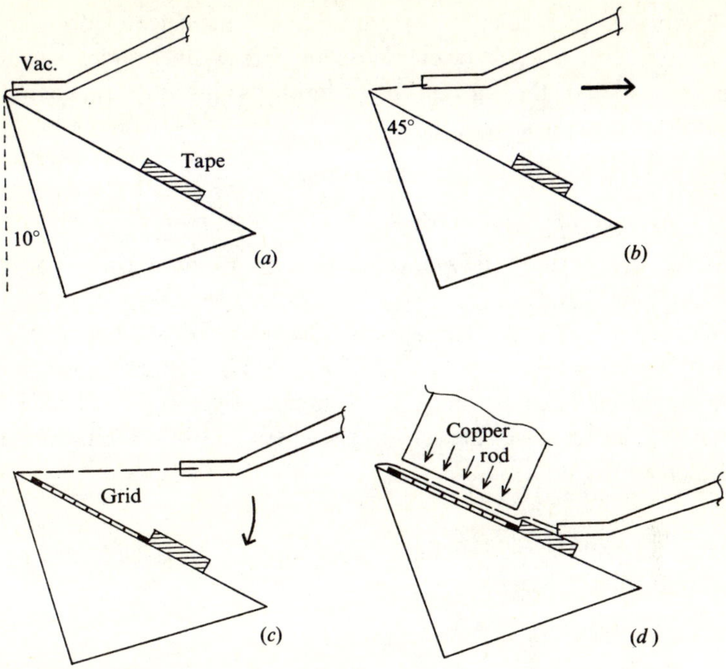

Fig. 2. Diagram illustrating method of pulling ribbons of ultrathin frozen sections onto dry glass knife. A weak vacuum, delivered through the flattened needle (Vac.) pulls the sections away from the knife edge. The vacuum device is moved back progressively as each section is cut until the ribbon is long enough to cover a grid. The Formvar-coated grid can be placed in front of the tape shelf either before sectioning commences or can be slipped underneath sections after the sections have been cut. The ribbon of sections is lowered over the grid as shown in (*c*) and the sections pressed lightly into contact with the grid using the polished copper rod (as in *d*). The grids can then be removed for freeze-drying, etc.

Freeze-drying

A cold stage and transfer system have been built which will enable frozen sections to be viewed and analysed in the electron microscope (AEI; EMMA-4) at −140 °C. Preliminary results have shown the desirability of keeping frozen sections in the hydrated state (Echlin, Paden, Dronzek & Wayte, 1970; Echlin, 1971). A detailed evaluation of the microscopy and analysis of hydrated ultrathin frozen sections will be published elsewhere.

The results shown in this paper are from *freeze-dried* frozen sections viewed and analysed at either room temperature (Figs. 2–14 and 16–18) or at −140 °C (Fig. 15 and Table 4). Sections were freeze-dried for a minimum time of 3 h at −85 °C in the dry atmosphere of the cryostat: they were then slowly warmed to ambient temperature in the presence of dry nitrogen gas, and coated with a thick layer of carbon (10–30 nm) under vacuum. Tests have shown that the carbon layer protects the dry sections from the humidity of laboratory air.

Microscopy and analysis

Unstained freeze-dried frozen sections were examined in either the AEI EMMA-4, or JEOL 100C electron microscopes. X-ray microanalysis of selected areas was carried out using wavelength dispersion (EMMA-4) or energy dispersion (EMMA-4 and JEOL 100C; both microscopes were equipped with Kevex detectors and Link System analysers). Details of both forms of X-ray microanalysis are given elsewhere in this volume (chapters by Chandler and Hutchinson).

Results

Frozen sections

Attempts have been made to test the reliability of frozen sections. Sodium chloride was dissolved in a solution of Tissue-Tec (carboxymethyl cellulose) at a concentration of 1 M: at this concentration the 'solution' showed a distinct floccular appearance. The 'solution' was frozen, sectioned at 100 nm, and either freeze-dried as described, or allowed to melt in a jet of dry nitrogen. The results are shown in Figs. 3 and 4 and in Table 2. In the freeze-dried preparations no sodium was detectable in the clear areas between the feather-like structures. In the melt-dried preparations 'clear' areas between 'snowflakes' showed significant concentrations of sodium. These results suggest that in freeze-dried preparations there was no diffusion of sodium from the sodium-rich feathers into the clear spaces between; the results for the melt-dried samples show that diffusion did occur and could be detected.

The potassium/sodium ratio in human red blood cells determined by X-ray microanalysis of freeze-dried frozen sections (100 nm) was 8.45 ± 1.37: the expected ratio (Ruch & Patton, 1966) is 7.35. The small difference between the actual and expected [K]/[Na] could be due to (*a*) the low counting

---

Fig. 4. Electron micrograph of similar material to Fig. 3 but after melt drying. Note the difference in structure showing spread of crystals away from dense areas. All 'clear' areas showed significant sodium. (After Appleton, 1974*a*.)

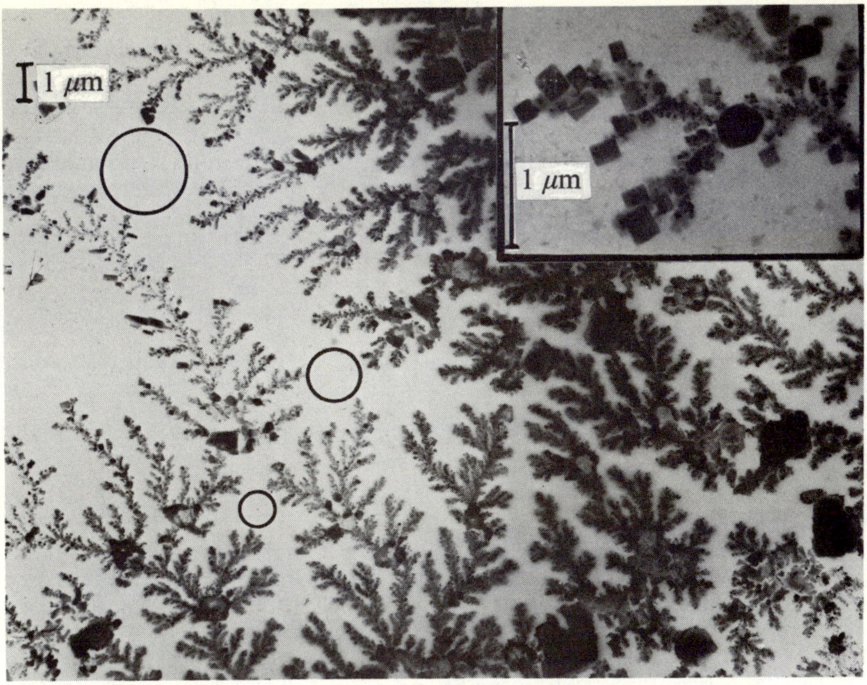

Fig. 3. Electron micrograph of section of carboxymethyl cellulose containing 1 M sodium chloride. Elaborate feather-like structures with high sodium content can be seen. Circles indicate typical areas of analysis (referred to in Table 2). Insert shows enlargement of the tip of the feathers; individual sodium chloride crystals can be seen. Preparation was freeze-dried. (After Appleton, 1974*a*.)

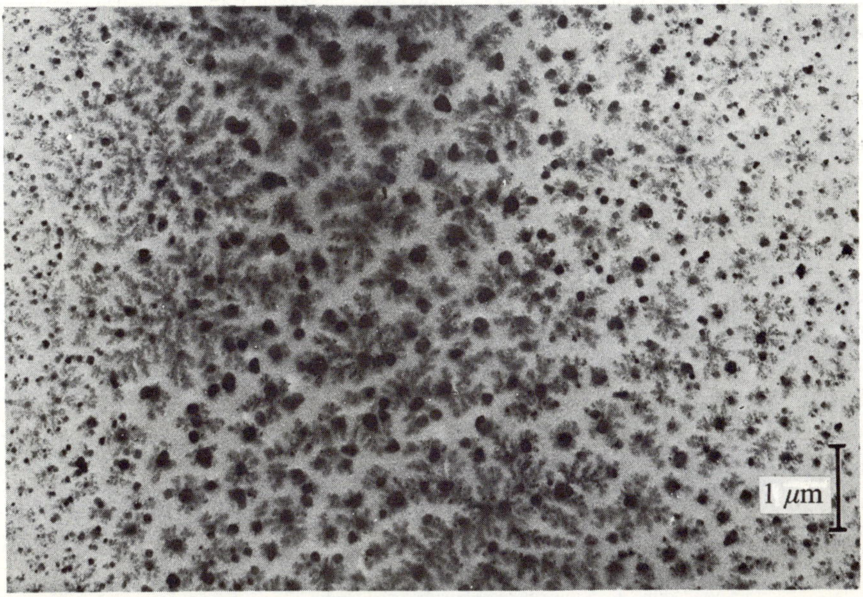

Table 2. *Analysis of sodium in sodium chloride/carboxymethyl cellulose sections illustrated in Figs. 3 and 4*

	Area	Peak	Background	Peak: background ratio
Freeze-dried	Over feathers (dense)	838 ± 249	52 ± 11	10:1
	Over fine feathers	417 ± 105	27 ± 1	10:1
	Just touching fine feathers	136 ± 20	15 ± 1	7:1
	Just not touching feathers	17 ± 15	9 ± 2	Not significant
Melt-dried	Over dense parts	657 ± 231	23 ± 2	18:1
	Over fine particles*	98 ± 36	18 ± 8	3:1

Concentration of sodium chloride was 1 M.

Background due to scattering was higher in the freeze-dried specimens due to the presence of more well defined dense deposits which result in scattering of stray electrons.

* Areas counted over fine particles in melt-dried samples (Fig. 12) were always more than 10 μm from the dense parts.

efficiency for weak sodium X-rays and (*b*) greater movement or volatilisation of sodium during electron bombardment.

Further evidence supporting these results is given in Table 3. Four naturally occurring elements were analysed in a cell of mouse pancreas after freeze-drying. The results suggest that intracellular compartments are retained even when the separation between areas (e.g. A and B) was only 100 nm. One would not expect to get such marked differences in elemental content if diffusion due to thawing during sectioning was significant (Appleton, 1972*a*, *b*, 1974*a*, *b*).

Morphology

The morphology in freeze-dried frozen sections of unfixed, unembedded and unstained biological material is often very different from that obtained after conventional fixation, dehydration, embedding and staining: membranes are not always seen in transmission electron micrographs of freeze-dried frozen sections probably because the material within the membranes has the same electron density as the material on either side of those membranes (for a detailed discussion of the morphology see Appleton, 1974*a*): Figs. 5, 7, 8 and 16 show transmission electron micrographs of ultrathin frozen sections.

Table 3. *Analysis of elements in an exocrine cell of mouse pancreas* (*Appleton 1972a, b*)

Area	Sodium	Potassium	Phosphorus	Calcium
(A) Nucleus (uncondensed)	1.44	41.40	59.00	—
(B) Nucleus (condensed)	5.30	123.00	117.20	2.72
(C) Endoplasmic reticulum (rough)	3.20	97.00	118.80	2.26
(D) Ground cytoplasm	4.76	151.00	121.00	—
(E) Secretory vesicles	3.25	122.30	45.80	4.26

Analysis of each element for an area is given by:

$$\frac{\text{the characteristic count for the element}}{\text{White count}} \times 100.$$

White count is given by a non-dispersive detector and is used as a measure of mass; changes in section thickness have therefore been corrected for.
Up to four locations were counted for each area.
Spot, or probe, size was 150 nm.

However membrane structures can be seen when the surface of a freeze-dried section is viewed by scanning electron microscopy (SEM; Fig. 6) because the various components of the cell freeze-dry down to different levels. This is essentially an artifact which can be used to confirm the identification of organelles as seen in transmission electron micrographs (TEM) or scanning transmission electron micrographs (STEM; Fig. 5).

Membrane boundaries are sometimes seen as an array of granules (Fig. 8); these granules may represent part of the subunit structure of membranes (Appleton 1974*a*).

X-ray microanalysis of biological material

Extruded axoplasm from the giant axon of the fan worm *Myxicola infundibulum** was frozen and sectioned as described. Sections were either freeze-dried (as described), freeze-substituted in acetone at −75 °C overnight, or melt-dried. X-ray microanalysis of the mitochondria and the immediately adjacent axoplasm (0.25 μm from mitochondria) was carried out in EMMA-4 using energy dispersive analysis and the results are shown in Figs. 9–11. These spectra were compared with the spectra obtained after analysing mitochondria in extruded axoplasm after glutaraldedyde fixation (in the presence of 1 mM Ca^{2+}), dehydration and embedding in araldite

* The *Myxicola* material was provided by Dr W. W. Schlaepfer and Professor P. F. Baker.

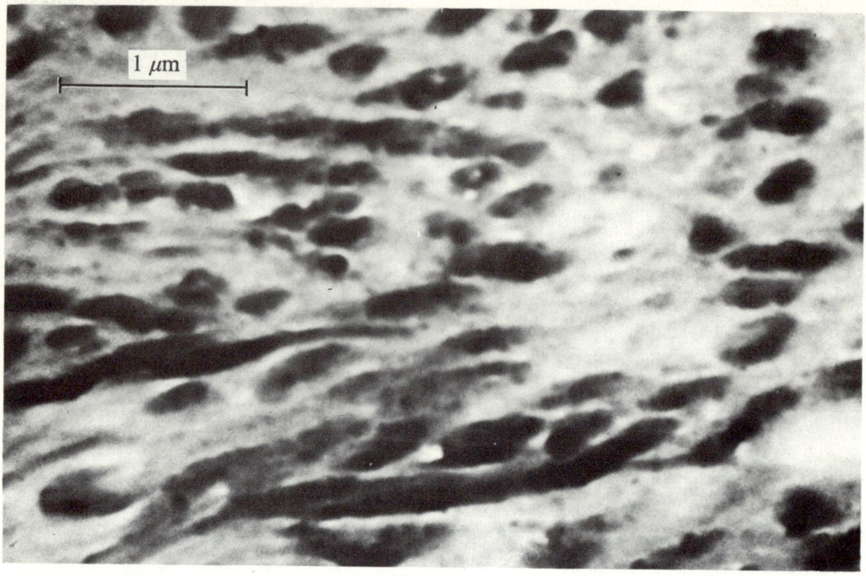

Fig. 5. Scanning transmission electron micrograph (STEM) of mitochondria in convoluted proximal tubule of mouse kidney: 100 nm frozen section freeze-dried. Note that little detail is visible in the dense structures. (Jeol 100C; unfixed, unstained; accelerating voltage 80 kV.)

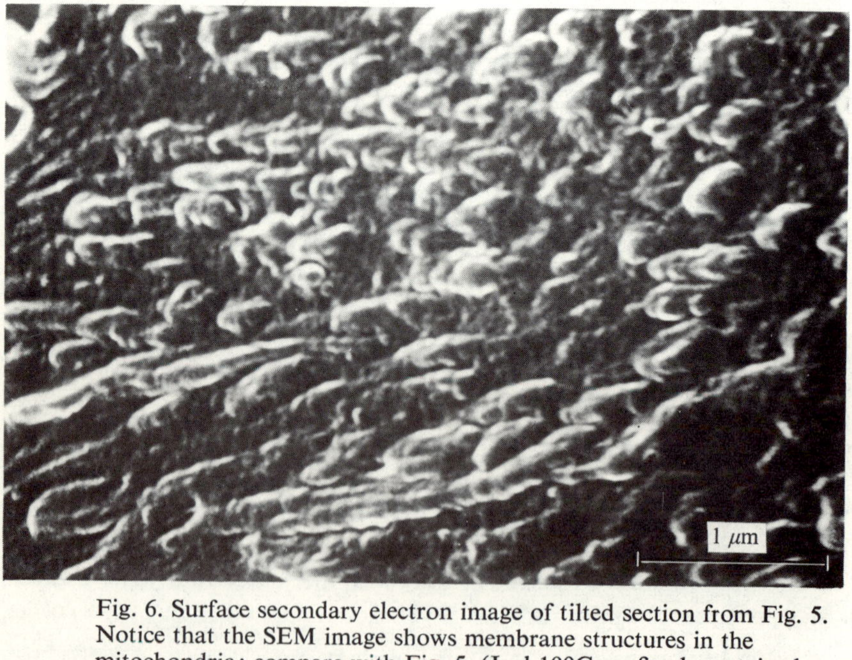

Fig. 6. Surface secondary electron image of tilted section from Fig. 5. Notice that the SEM image shows membrane structures in the mitochondria; compare with Fig. 5. (Jeol 100C; unfixed, unstained; accelerating voltage 80 kV.)

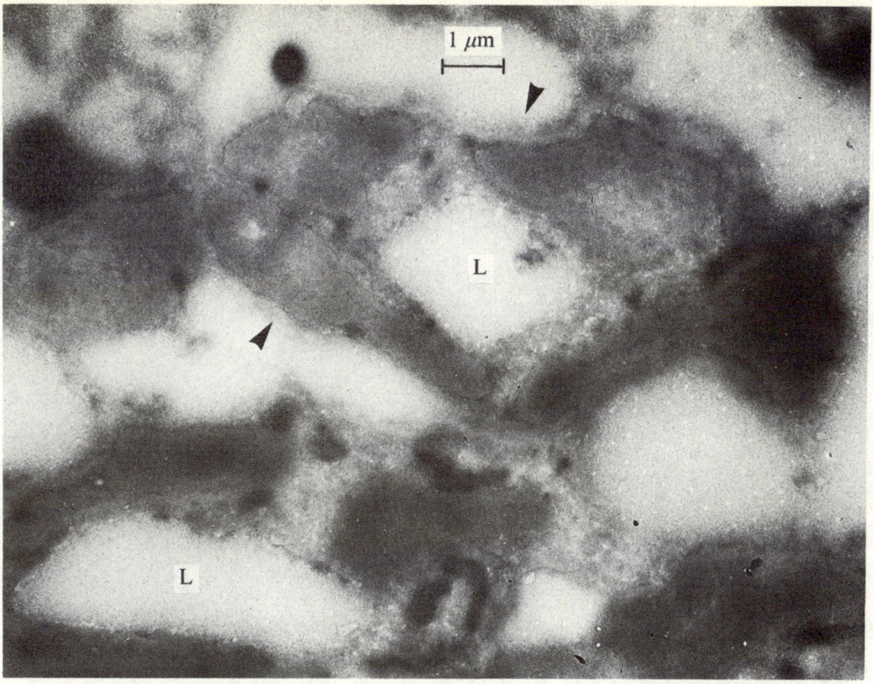

Fig. 7. Transmission electron micrograph (TEM) of small blood vessel in mouse pancreas. L, lumen; arrows show basement membranes. (EMMA-4; unfixed, unstained.)

(Fig. 12). Considerable loss or redistribution of elements occurs in the freeze-substituted, melt-dried and glutaraldehyde-fixed material.

In another experiment extruded axoplasm was dialysed against a solution containing 500 mM potassium glutamate, 5 mM potassium phosphate, 5 mM ATP, and 2 mM calcium acetate; the dialysed axoplasm was frozen, sectioned and freeze-dried as described. Figs. 13 and 14 show the X-ray spectra when the sections were analysed at room temperature (+18 °C). Fig. 15 shows the effect of analysis at −140 °C; the significance of low temperature analysis will be discussed below.

X-ray microanalysis has shown that the ionic concentrations can vary from one part of a cell to another even when the separation of those areas is as small as 100 nm (Table 3) or 250 nm (Figs. 13 and 14 and Table 4). Other experiments have shown that neighbouring cells may also have different ionic concentrations (Figs. 16, 17 and 18).

X-ray microanalysis has also shown that the elemental composition can be altered experimentally – that mitochondria in the giant axon of *Myxicola* can take up calcium when dialysed with Ca^{2+} and ATP (compare Figs. 11 and 13). However, since the volume occupied by the mitochondria is only

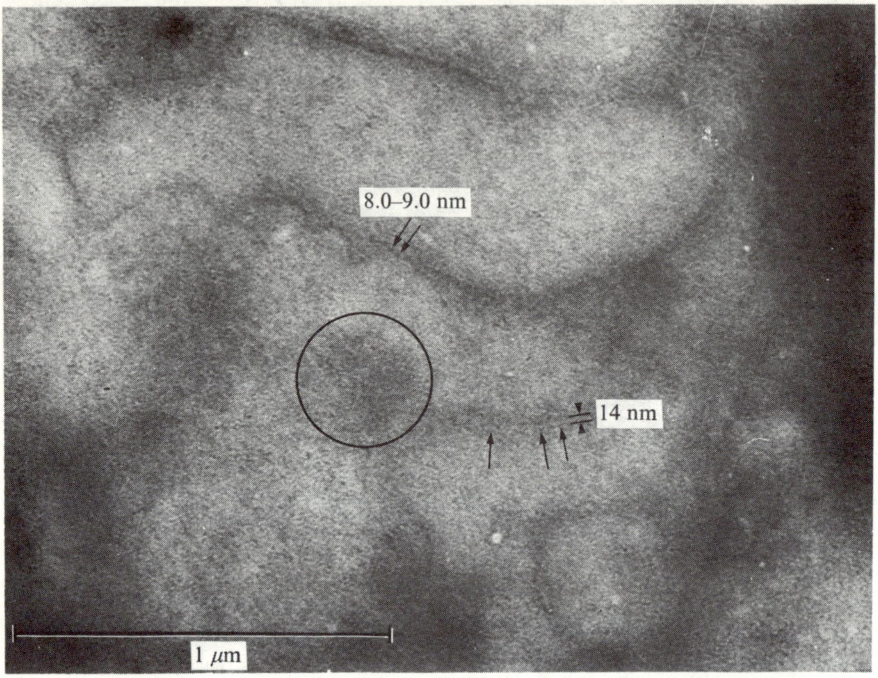

Fig. 8. Membrane granulation in Golgi region of mouse pancreas, unfixed and unstained. Size of granules is 8.04 ± 0.94 nm. A length of double array (arrows) shows a separation of 14 nm. Granules show centre-to-centre separation of about 14 nm. Area within circle is probably an area where the plane of sectioning runs parallel to surface of membrane and shows a periodicity of 14 nm. (After Appleton 1974*a*.)

about 1 % of the total axoplasmic volume, it is doubtful whether calcium storage by the mitochondria is sufficient to account for the needs of the axon. This work is to be extended using the squid giant axon.

X-ray microanalysis at low temperatures

The cold stage developed for the EMMA-4 electron microscope makes it possible to analyse specimens at −140 °C. At these temperatures the characteristic peaks for calcium and potassium are increased and the overall background is reduced (Fig. 15 and Table 4).

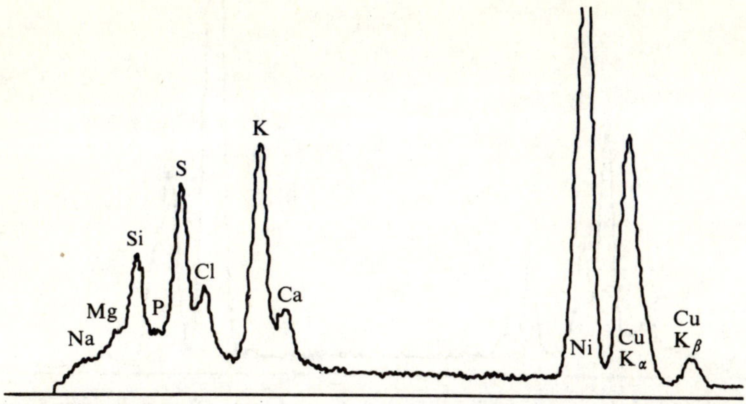

Fig. 9. Energy dispersive X-ray spectrum from control mitochondrion of extruded axoplasm from giant axon of *Myxicola infundibulum.* 100-nm frozen section unfixed, unstained, freeze-dried. Counting time, 100 s; vertical gain, 4. In comparing analytical spectra note that a vertical gain of '4' represents a vertical 'enlargement' twice that of vertical gain '3' and vice versa. All spectra were obtained with a beam current of 4 nA and at an accelerating voltage of 60 kV. The peaks of cobalt, nickel and copper are background due to grid material and copper in specimen rods.

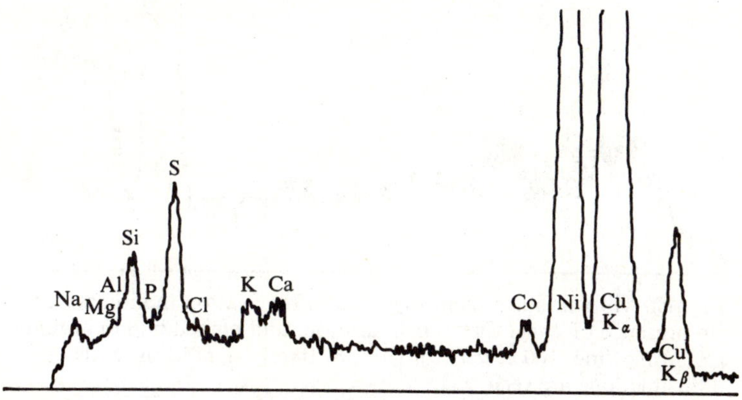

Fig. 10. Mitochondria as in Fig. 9 but after freeze-substitution in acetone at −75 °C. Counting time, 200 s; vertical gain, 6. Note that in spite of the increased vertical gain (4 times that of Fig. 9) and increased counting time potassium and chlorine are virtually gone and calcium is greatly reduced.

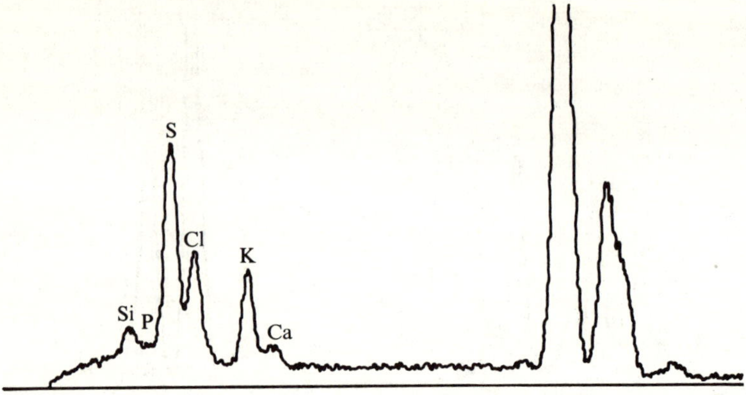

Fig. 11. Mitochondria from *Myxicola* axoplasm after melt-drying. Counting time, 200 s; vertical gain, 5. Compare with Fig. 9. Taking into account increased counting time and increased vertical gain (×2) potassium, chlorine, calcium and magnesium have been lost (i.e. diffused during melting).

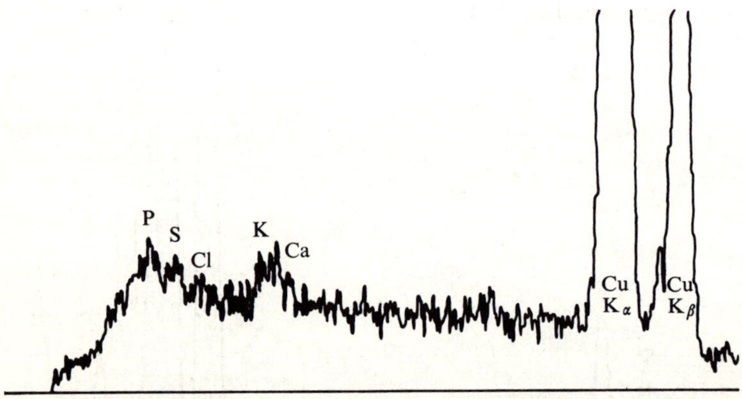

Fig. 12. Mitochondria from *Myxicola* after fixation in glutaraldehyde (in presence of 1 mM Ca^{2+}), dehydration, and embedding in araldite. Counting time, 100 s; vertical gain, 8. Barely significant levels of diffusible ions are seen.

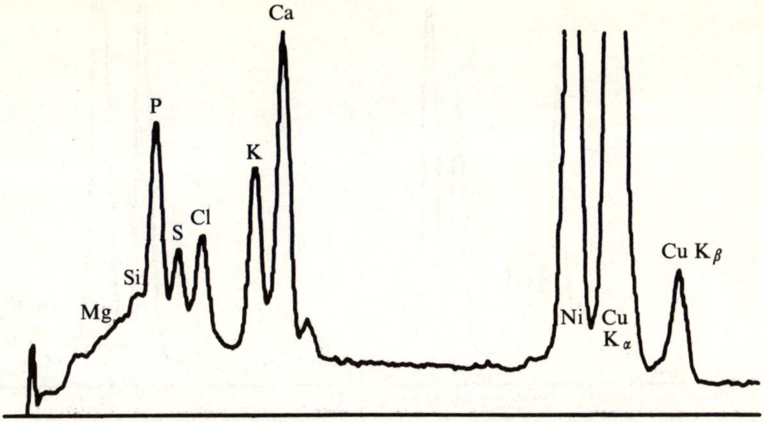

Fig. 13. Mitochondria from extruded axoplasm of *Myxicola* dialysed against 2 mM calcium acetate; 5 mM ATP; 5 mM potassium phosphate; 500 mM potassium glutamate. Two hours' dialysis at 37 °C; counting time, 200 s; vertical gain, 5; analysed at +18 °C. Note the uptake of calcium (compare with controls in Fig. 9 and see Table 5).

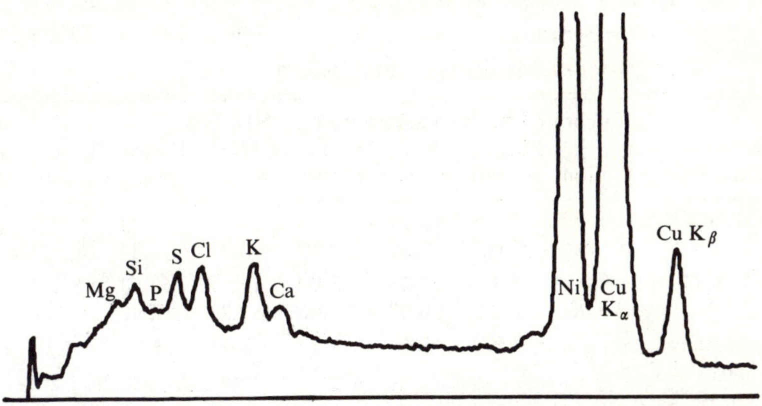

Fig. 14. X-ray spectrum from immediately adjacent cytoplasm to mitochondrion in Fig. 13. Counting time, 200 s; vertical gain, 5; analysed at +18 °C. Treatment with Ca^{2+}/ATP (as in Fig. 13) has little effect on adjacent cytoplasm. All the calcium appears to go into the mitochondria. (See also Table 5.)

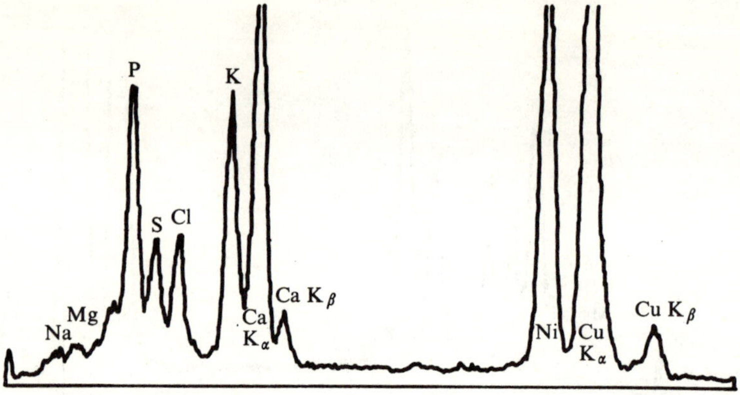

Fig. 15. X-ray spectrum from adjacent Ca^{2+}/ATP treated mitochondria (1 μm away from mitochondrion analysed in Fig. 13). Counting time, 200 s; vertical gain, 5; analysed at −140 °C. Note that the potassium and calcium peaks are higher and the background lower than in Fig. 13. (See also Table 4.)

Table 4. *The effects of analysing freeze-dried frozen sections at low temperatures: mitochondria from Ca^{2+}/ATP treated extruded axoplasm of* Myxicola infundibulum

Temperature	(Integrated peak − background ÷ W) × 100 Sodium	Magnesium	Phosphorus	Sulphur	Chlorine	Potassium	Calcium	Continuum (W)
+18 °C	—	0.130 ±0.07	10.23 ±2.38	4.40 ±0.17	4.53 ±0.22	17.93 ±1.21	12.23 ±2.25	36232 ±956
−140 °C	0.070 ±0.02	1.150 ±0.22	55.43 ±0.02	13.90 ±0.66	16.07 ±1.23	62.53 ±2.96	52.27 ±1.38	11845 ±1540
Increase in sensitivity at −140 °C	[a]	8.9	5.4	3.2	3.6	3.5	4.3	0.33[b]

[a] No figure can be given for increased sensitivity for sodium at −140 °C until further data have been acquired.

[b] The continuum (W) is used as a measure of mass and is taken from a 'window' set in the spectrum between 13.00 and 15.00 keV energies. The continuum is reduced at −140 °C due to lowering of contamination (see Discussion and conclusions' section).

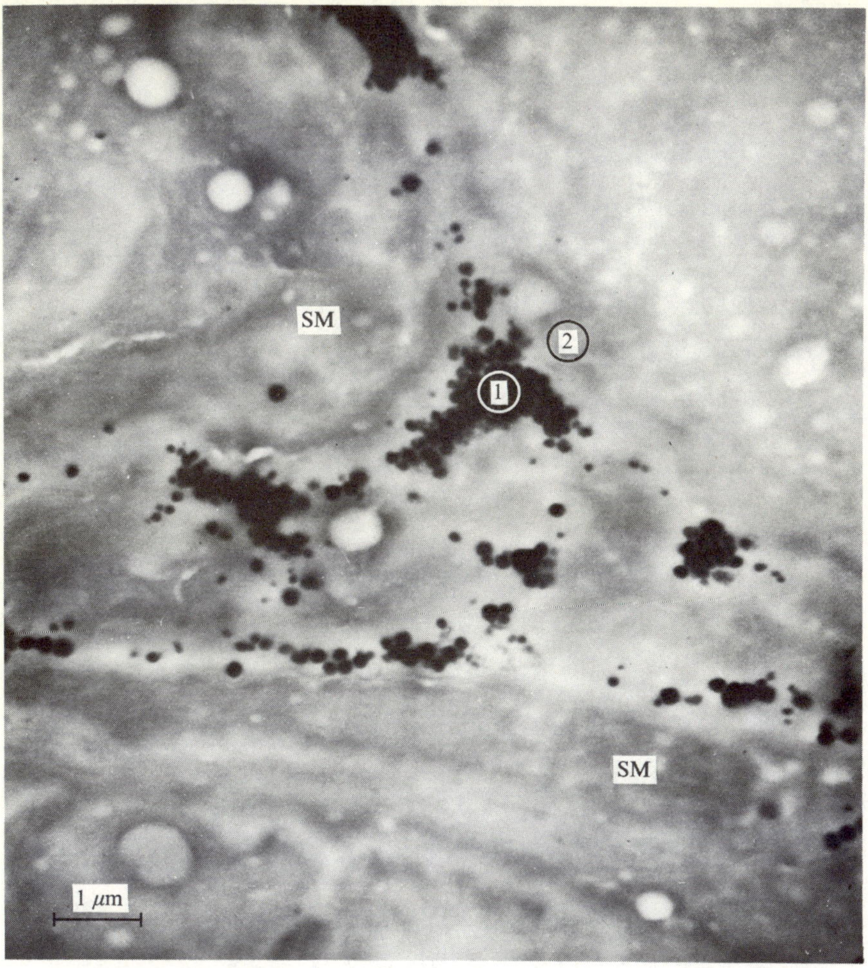

Fig. 16. Scanning transmission electron micrograph (STEM) of frozen section unfixed and unstained of parasitic helminth worm (*Polymorphus minutus*). Note granular (possibly ganglion) cells lying between smooth muscle (SM) cells. Area (1) analytical spectrum shown in Fig. 17; area (2) analytical spectrum shown in Fig. 18.

Quantification

The elemental concentrations obtained by X-ray microanalysis can be quantified in terms of percentage mass fractions by comparing the characteristic counts for the elements in the specimen with those from standards of known composition (see Chandler, this volume).

The percentage mass fractions of calcium present in (*a*) control mito-

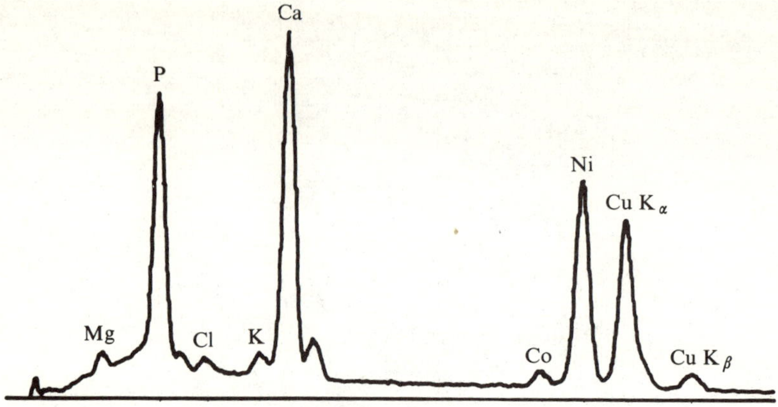

Fig. 17. X-ray analytical spectrum over area (1) from Fig. 16. Counting time, 200 s; vertical gain, 3.

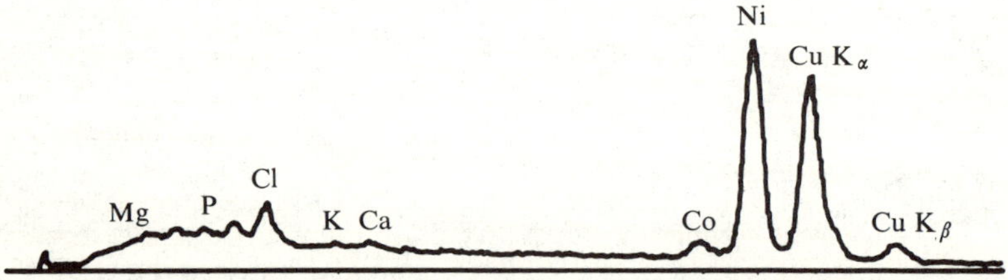

Fig. 18. Spectrum obtained by analysing area (2) from Fig. 16. Counting time, 200 s; vertical gain, 3. Note the very different spectra obtained after analysing two neighbouring cells.

chondria of *Myxicola* (e.g. Fig. 9), (*b*) adjacent cytoplasm, (*c*) mitochondria of *Myxicola* treated with Ca^{2+}/ATP (Fig. 13), and (*d*) adjacent cytoplasm of treated axoplasm, are given in Table 5. The data were quantified according to the method described by Hall, Anderson & Appleton (1973) using a fine separation of calcium chloride crystals as a standard.

The results shown in Table 5 are expressed as percentage mass fractions of the *dry* weight. To convert such figures to millimolar concentrations is only possible if the water content of the mitochondria and adjacent cytoplasm is

Table 5. *Quantified results from X-ray microanalysis of frozen sections of extruded axoplasm from the giant axon of* Myxicola infundibulum

Treatment	Area		% mass fraction (Ca)	Mean ± S.E.
Ca^{2+}/ATP[a]	Mitochondria	1	2.36	
		2	2.09	2.38 ± 0.34
		3	1.75	
		4	3.31	
	Adjacent	1	0.09	
	cytoplasm[b]	2	0.14	0.16 ± 0.02
		3	0.20	
		4	0.22	
Controls[c]	Mitochondria	1	1.00	
		2	0.34	
		3	0.33	
		4	2.71	
		5	2.45	1.05 ± 0.32
		6	0.17	
		7	0.58	
		8	0.88	
	Adjacent	1	1.19	
	cytoplasm[b]	2	0.23	
		3	0.26	
		4	0.13	0.35 ± 0.16
		5	0.25	
		6	0.25	

[a] Extruded axoplasm from giant axon of *Myxicola infundibulum* dialysed for 2 h against 500 mM potassium glutamate: 5 mM potassium phosphate; 5 mM ATP; 2 mM calcium acetate.
[b] Adjacent cytoplasm was always 0.25 μm away from mitochondria.
[c] Extruded axoplasm frozen immediately after extrusion.

known. The *estimated* millimolar concentrations for the mass fraction data are given in Table 6 based on the *assumption* that the water content (in the living state) of the mitochondria is 90 % and that for adjacent cytoplasm 95 %.

Discussion and conclusions

One of the criticisms which is often levelled against the use of frozen sections is that unless the temperature of sectioning is below −100 °C diffusible substances could be redistributed due to momentary thawing at the knife edge. Thornburg & Mengers (1957) suggested that the heat dissipated per centi-

Table 6. *% mass fractions from Table 5 converted to millimolar concentrations assuming 90 % water content for mitochondria and 95 % for adjacent cytoplasm*

Treatment	Area	% mass fraction Ca (dry weight)	mM l^{-1} Ca (hydrated)
Ca^{2+}/ATP	Mitochondria	2.38	66
	Cytoplasm	0.16	4.4
Controls	Mitochondria	1.05	14
	Cytoplasm	0.35	4.6

metre of knife travel could be 1.96×10^4 erg and that, allowing for the latent heat of fusion of ice, this would melt a layer of 60 nm in a section 100 nm thick. Hodson & Marshall (1970, 1972) have calculated that the temperature rise at the point of sectioning was as high as 100 deg C.

The results with the sodium chloride/carboxymethyl cellulose 'solution' (Table 2) and the evidence for elemental compartments in the pancreas cell (Table 3) suggest that there is no redistribution due to momentary thawing at the knife edge, and that significant differences in elemental concentration can be detected within a resolution of 100 nm. Results from X-ray microanalysis of mitochondria and adjacent cytoplasm of the axon of *Myxicola infundibulum* confirm these findings. The sensitivity of X-ray microanalysis using freeze-dried frozen sections would appear to be as low as 0.5 mM and may be substantially increased by analysing frozen sections at −140 °C.

The apparent increase in sensitivity at low temperatures is probably due to two effects: (*a*) the stability of some elements is greater at −140 °C (particularly potassium and calcium; see Fig. 15); and (*b*) the greater peak-to-background ratios due to a marked lowering of the overall background. The lowering of the background is probably a contamination effect. The contamination rate in the EMMA-4 cold stage appears much lower than that at room temperatures. Furthermore it is known (Ruska, 1966) that carbon contamination is 'stripped' from specimens when the temperature of the stage is reduced below −80 °C. No reduction in mass has so far been observed when freeze-dried frozen sections are viewed at −140 °C in EMMA-4 (80 kV accelerating voltage; 4 nA beam current). It seems, therefore, that it is the carbon contamination which is being removed during analysis rather than carbon within the section. Further tests are being conducted to establish the nature of the improved sensitivity of X-ray microanalysis at low temperatures. The results of these tests and the effects of analysing hydrated ultrathin frozen sections will be reported elsewhere.

Baker & Appleton (1976) have recently shown that ultrathin frozen sections containing radioactive compounds and coated with carbon can be used for autoradiography of diffusible substances. X-ray microanalysis has shown that the layer of carbon over the sections protects the diffusible substances from moisture in the overlying emulsion. Both techniques of X-ray microanalysis and autoradiography can therefore be used on the same material. Autoradiography can be used to localise radioactive drugs and toxic agents; X-ray microanalysis can be used to see how the chemistry of the cell is changed by those substances. The potential information which can now be obtained is therefore very great (Appleton, Shute & Baker, 1976). Biologists are now able to look right into the cell in order to study its chemistry under varying experimental conditions and at the ultrastructural level.

References

Appleton, T. C. (1968). The application of autoradiography to the study of soluble compounds. *Acta Histochemica, Supplement*, **8**, 115–33.

Appleton, T. C. (1969). The possibilities of locating soluble labelled compounds by electron microscope autoradiography. In *Autoradiography of Diffusible Compounds*, ed. W. E. Stumpf & L. J. Roth, pp. 301–19. Academic Press, New York & London.

Appleton, T. C. (1972*a*). 'Dry' ultrathin frozen sections for electron microscopy and X-ray microanalysis; the cryostat approach. *Micron*, **3**, 101–5.

Appleton, T. C. (1972*b*). 'Dry' ultrathin frozen sections of unfixed and unembedded biological material for X-ray microanalysis of naturally occurring diffusible electrolytes; the cryostat approach. *Journal de Microscopie*, **13**, 144–6.

Appleton, T. C. (1974*a*). A cryostat approach to ultrathin 'dry' frozen sections for electron microscopy; a morphological and X-ray analytical study. *Journal of Microscopy*, **100**, 49–74.

Appleton, T. C. (1974*b*). Cryoultramicrotomy; possible applications in cytochemistry. In *Electron Microscopy and Cytochemistry*, ed. E. Wisse, W. Th. Daems, I. Molenaar & P. van Duijn, pp. 229–41. North-Holland Publishing Co., Amsterdam.

Appleton, T. C., Shute, C. C. D. & Baker, J. R. J. (1976). Towards an understanding of toxicity. *Trends in Biochemical Sciences*, in press.

Baker, J. R. J. & Appleton, T. C. (1976). A technique for electron microscope autoradiography (and X-ray microanalysis) of diffusible substances using freeze-dried fresh frozen sections. *Journal of Microscopy*, **108**, 307–15.

Bernhard, W. (1971). Improved techniques for the preparation of ultrathin frozen sections. *Journal of Cell Biology*, **49**, 731–46.

Christensen, A. K. (1969). A way to prepare thin frozen sections of tissue for electron microscopy. In *Autogradiography of Diffusible Substances*, ed. W. E. Stumpf & L. J. Roth, pp. 349–62. Academic Press, New York & London.

Christensen, A. K. (1971). Frozen thin sections of fresh tissue for electron microscopy, with a description of pancreas and liver. *Journal of Cell Biology*, **51**, 772–804.

Echlin, P. (1971). The examination of biological material at low temperatures. In *Proceedings of the 4th Scanning Electron Microscope Symposium*, ed. by O. Johari, pp. 225–32. IIT Research Institute, Chicago.

Echlin, P., Paden, R., Dronzek, B. & Wayte, R. (1970). Scanning electron microscopy of labile biological material maintained under controlled conditions. In *Proceedings of the 3rd Scanning Electron Microscope Symposium*, ed. O. Johari, pp. 49–56. IIT Research Institute, Chicago.

Hall, T. A., Anderson, H. C. & Appleton, T. C. (1973). The use of thin specimens for X-ray microanalysis in biology. *Journal of Microscopy*, **99**, 177–82.

Hodson, S. & Marshall, J. (1970). Ultramicrotomy; a technique for cutting ultrathin frozen sections of unfixed biological tissues for electron microscopy. *Journal of Microscopy*, **91**, 105–17.

Hodson, S. & Marshall, J. (1972). Evidence against through-section thawing whilst cutting on the ultracryotome. *Journal of Microscopy*, **95**, 459–66.

Ruch, T. C. & Patton, H. D. (1966). *Physiology and Biophysics*. W. B. Saunders & Co., Philadelphia.

Ruska, E. (1966). Past and present attempts to attain the resolution limit of the transmission electron microscope. In *Advances in Optical and Electron Microscopy*, ed. R. Barer & V. E. Cosslett, vol. 1, pp. 115–79. Academic Press, New York & London.

Sjöström, M. & Thornell, L. E. (1975). Preparing sections of skeletal muscle for transmission electron analytical microscopy (TEAM) of diffusible elements. *Journal of Microscopy*, **103**, 101–12.

Sjöström, M., Thornell, L. E. & Cedergren, E. (1973). The application of cryo-ultramicrotomy in the study of fine structure of myofilaments. *Journal of Microscopy*, **99**, 193–204.

Thornburg, W. & Mengers, P. E. (1957). An analysis of frozen section techniques; sectioning of fresh-frozen tissue. *Journal of Histochemistry and Cytochemistry*, **5**, 47–52.

INDEX